X-Ray Metrology in Semiconductor Manufacturing

X-Ray Metrology in Semiconductor Manufacturing

D. Keith Bowen
Brian K. Tanner

Taylor & Francis
Taylor & Francis Group
Boca Raton London New York

A CRC title, part of the Taylor & Francis imprint, a member of the Taylor & Francis Group, the academic division of T&F Informa plc.

Published in 2006 by
CRC Press
Taylor & Francis Group
6000 Broken Sound Parkway NW, Suite 300
Boca Raton, FL 33487-2742

Printed in the United States of America on acid-free paper
10 9 8 7 6 5 4 3 2 1

International Standard Book Number-10: 0-8493-3928-6 (Hardcover)
International Standard Book Number-13: 978-0-8493-3928-8 (Hardcover)
Library of Congress Card Number 2005052196

Library of Congress Cataloging-in-Publication Data

Bowen, D. Keith (David Keith), 1940-
X-ray metrology in semiconductor manufacturing / by David K. Bowen, Brian K. Tanner.
p. cm.
Includes bibliographical references and index.
ISBN 0-8493-3928-6 (alk. paper)
1. Semiconductors--Design and construction--Quality control. 2. Integrated circuits--Measurement. 3. Semiconductor wafers--Inspection. 4. X-rays--Diffraction. 5. Fluoroscopy. I. Tanner, B. K. (Brian Keith) II. Title.

TK7874.58.B69 2006
621.3815'2--dc22 2005052196

informa
Taylor & Francis Group
is the Academic Division of Informa plc.

Visit the Taylor & Francis Web site at
http://www.taylorandfrancis.com

and the CRC Press Web site at
http://www.crcpress.com

Preface

Semiconductor manufacturing technology in recent years has evolved to the point at which traditional metrologies, largely based upon optical techniques, are no longer adequate for process development or product monitoring. Thin films in current manufacturing processes may be less than 1 nm in thickness. Measuring these with a tool whose probing wavelength is several hundred nanometers is akin to measuring the thickness of a pencil line with a yardstick. X-rays, with their wavelengths around 0.1 nm, are clearly appropriate. Moreover, the constants that describe their interaction with materials are "Goldilocks" values, just right for the requirement. X-ray metrology (XRM) is now being rapidly adopted in manufacturing industry for semiconductor, magnetic, and other advanced thin-film materials.

This book is about wafer metrology for the semiconductor industry by x-ray methods. Its scope includes the highly accurate and traceable interferometric methods of diffraction and specular reflectivity, the methods of diffuse scatter that require detailed modeling but provide unique insights into film structure, and the simpler intensity methods, such as x-ray fluorescence, which require calibration but give valuable complementary information about material composition and mass density.

The metrologies that ensue are appropriate for measuring film thickness, composition, strain and its relaxation, crystallinity, mosaic spread, surface and interface roughness, and porosity and pore size. The techniques that have evolved include x-ray reflectivity (both specular and diffuse), high-resolution x-ray diffraction, diffraction imaging and interferometry, and fluorescence. The scope of this book is their application to measuring these parameters repeatably, accurately, and rapidly on development and production wafers.

Part 1 of the book, "The Applications," is intended to answer the following questions: Can I use x-rays to measure this parameter in my wafers? What are the limits of measurement? The key elements of the techniques are given by means of inset boxes in this part, which is organized by the parameters to be measured rather than by technique.

Part 2 of the book, "The Science," discusses the techniques and the basic theory underlying each. This is intended for the more specialized engineer or tool owner who wants to see whether a particular technique is well established in theory or is more speculative, or wants to discover whether it can be pushed to solve new materials problems. The theory is described and assessed, but detailed derivations are not given, since they are readily available in earlier publications.

Part 3 of the book, "The Technology," deals with the practical implementation of x-ray metrology. First, the technique of automated data analysis and modeling is covered, followed by the instrumentation fundamentals for the various techniques. Topics such as x-ray optics are discussed in terms of their contribution and potential to solve metrological problems, such as sufficient intensity in a small spot, rather than in academic detail. The concluding chapter covers the essential metrological questions of precision and repeatability, absolute accuracy, spot size, and throughput for each type of measurement.

Acknowledgments

It is a pleasure to acknowledge the assistance we have had from many colleagues in the preparation of this book.

We first thank those colleagues, and their publishers, who have given us permission to use their published figures and data. These are acknowledged individually in the figure captions.

This book contains a great number of previously unpublished figures from colleagues at Bede X-ray Metrology and we warmly thank those colleagues who have taken the data and prepared the figures and tables. We also owe them thanks for innumerable technical and scientific discussions over the years, which have greatly contributed to our own understanding of x-ray metrology. These are: Matthew Wormington, in particular, for figures and discussions on porosity, x-ray reflectivity and diffuse scatter, genetic algorithms, and data fitting; Paul Ryan, in particular, for figures, tables, and discussions on repeatability and reproducibility, SiGe diffraction, and x-ray fluorescence; Kevin Matney for many discussions and figures and, in particular, those on reciprocal space mapping and texture analysis; Tamzin Lafford for many discussions and figures on experimental measurements on x-ray diffraction and reflectivity throughout the book; Petra Feichtinger, for the discussions and all the experimental figures on x-ray diffraction imaging; David Joyce, for figures on x-ray reflectivity; Richard Bytheway for figures on x-ray fluorescence; and Ladislav Pina, Neil Loxley, and John Wall, for discussions and figures on x-ray sources and optics.

We likewise thank colleagues from the University of Durham for their similar cooperation, assistance, and discussions. They are: Dr. Tom Hase, who has been pivotal in the Durham high resolution scattering group for many years; Prof. Peter Hatton, whose individual approach to x-ray scattering is always stimulating; Drs. Ian Pape, Brian Fulthorpe, Andrea Li-Bassi, James Buchanan, Stuart Wilkins, Amir Rozatian, and Alex Pym and other research students who have borne the brunt of much experimental data collection.

We also thank Petra Feichtinger, David Joyce, Tamzin Lafford, Paul Ryan, and Matthew Wormington for reviewing the entire book and making critical comments on the manuscript. There comments were invaluable. Needless to say, any errors are the responsibility of the authors.

Many customers of Bede X-ray Metrology allowed us to use data from their development samples. We may not name them individually but here we express our gratitude, since this enabled us to provide information on x-ray metrology in the most advanced materials.

We wish to thank the Directors of Bede plc for permission to publish this book and in particular Dr. Neil Loxley, CEO, for every encouragement and cooperation in its writing and preparation.

Finally, we thank Nora Konopka and the editing and publication team at Taylor & Francis for their encouragement, cooperation, and skill in the production of this book.

About the Authors

Professor Keith Bowen, F.R.Eng., F.R.S., obtained his M.A. and D.Phil. in metallurgy at Oxford University, working on mechanical properties of metals. He then held academic positions at Warwick University from 1968 onward, culminating in his appointment as professor of engineering and director of the Center for Nanotechnology and Microengineering, which he held until 1997. He has held visiting professorships at Massachusetts Institute of Technology, University of Paris, and University of Denver. He is currently emeritus professor of engineering at Warwick University and visiting professor in physics at Durham University. He has authored over 130 publications on the theory and application of x-ray characterization techniques, theory of dislocations, x-ray interferometry, and ultraprecision engineering, including the book *High-Resolution X-Ray Diffraction and Topography* with Professor Brian Tanner. He joined Bede Scientific part-time in 1983, was engineering director from 1984 to 2000, was president of Bede Scientific, Inc. from 1995 to 2002, and was group director of technology from the flotation of Bede plc in 2000 until he retired in 2005. During this period he was responsible for the strategic development of science and technology in the Bede plc group of companies, for the development of the industry's first fully automated x-ray metrology tools, and for numerous inventions in x-ray technology, including the BedeScan™ method of digital x-ray diffraction imaging. Professor Bowen is a fellow of the Royal Society, fellow of the Royal Academy of Engineering, fellow of the Institute of Physics, and fellow of the Institute of Materials, Minerals and Mining.

Professor Brian Tanner moved to Durham in 1973 as a university lecturer, after holding a junior research fellowship at Linacre College, Oxford. Promoted to senior lecturer in 1983, reader in 1986, and professor in 1990, he served as head of the Physics Department from 1996 to 1999. From 1999 to 2000 he held a Sir James Knott Foundation Fellowship, and from 2000 to 2001 he was a Leverhulme research fellow. Since 2000, part of his time has been spent as director of the North East Centre for Scientific Enterprise. He has served on numerous research council committees and panels, and from 1998 to 2000 was chairman of a scientific review committee at the European Synchrotron Radiation Facility in Grenoble. In 1978 he co-founded a spin-off company, Bede Scientific Instruments Ltd., that floated on the London Stock Exchange in November 2000 as Bede plc. It is the largest spin-off company from the University of Durham, now employing about 150 people in the U.K., U.S., China, and Czech Republic. Professor Tanner is a nonexecutive director of Bede plc. He has published over 300 papers in refereed international scientific journals, written two books, co-authored a third, and edited three more. His research interests lie in understanding the relationship between magnetic, optical, and structural properties of advanced materials, making particular use of high-resolution x-ray scattering. He is a fellow of the Institute of Physics and a fellow of the Royal Society of Arts.

Professors Bowen and Tanner have between them over 80 years experience in x-ray analysis of materials, and have collaborated for over 25 years. Both are enthusiastic amateur musicians and claim that their long collaboration in science and industry began by playing a clarinet and piano duet at the NATO ASI conference on Characterization of Crystal Growth Defects by X-Ray Methods, which they jointly organized in Durham in 1979.

In 2005, their distinction and collaboration were recognized when they jointly received the biennial C.S. Barrett Award at the Denver X-Ray Conference for "seminal contributions to the theory, instrumentation and computerized analysis of x-ray scattering and x-ray reflectivity and for unceasing efforts in teaching and popularizing these topics" from the International Committee for Diffraction Data.

Brian Tanner, pianist. Keith Bowen, clarinetist. (Photograph courtesy of Ruth Tanner.)

Contents

Part 1 The Applications

Part 2 The Science

Part 3 The Technology

Part 1

The Applications

1

Introduction

1.1 Scope of X-ray Metrology (XRM)

This book is about wafer metrology for the semiconductor industry using x-ray methods. Its scope includes the highly accurate and traceable interferometric methods of diffraction and specular reflectivity, the methods of diffuse scatter that require detailed modeling but provide unique insights into film structure, and the simpler intensity methods, such as x-ray fluorescence, which require calibration but give valuable complementary information about material composition and mass density.

Interferometric methods work by splitting the x-ray wavefront at some discontinuity in the material and detecting the recombined component waves, normally by the intensity of scattering as a function of angle. They are analogous to optical interference, but no external optical element is required to split the wavefront; a natural feature such as an interface or a crystal plane is used. Such methods include specular x-ray reflectivity, diffuse scatter, high-resolution diffraction, x-ray topography, and x-ray interferometry. The interference of x-rays allows us to make three fundamental measurements, from which several others derive. These are *thickness*, from the interference fringes generated by rays reflected from pairs of interfaces, *strain*, and *tilt*, from the positions (and displacements) of Bragg diffraction peaks. The parameters that we can measure or infer include:

- Thickness
- Strain
- Composition
- Mosaic spread
- Lateral nanostructure dimensions, including critical dimension (CD) measurements
- Porosity

No 'golden wafer' or 'golden tool' is required with such methods to ensure accuracy or tool matching. These arise naturally from the *traceability* of the measurements to natural or international standards. Thickness and strain only require knowledge of the x-ray wavelength. Composition measurement additionally requires a calibration of the (nearly linear) relationship between lattice parameter and composition, and porosity requires knowledge or calibration of the mass density of the matrix material.

Next we have x-ray scattering phenomena, in which the wavefront is divided more or less continuously, without a sharp interface. Examples are small-angle scattering from pores or surface topology and scatter from small grains. The measurements are also of the intensity of scattering as a function of angle, but they require knowledge or inference of some distribution function in order to interpret the measurement. The value of the parameter derived depends on the details of the model used and thus is not intrinsically traceable. Nevertheless they are of major importance, and sometimes x-ray scattering is the only way to measure the parameter nondestructively. These include:

- Roughness amplitude of surfaces and buried interfaces
- Interdiffusion or intermixing length
- In-plane length scale of roughness
- Fractal dimensionality of the roughness
- Grain size
- Pore size distribution

Finally, we have methods based on the intensity of scatter at essentially a single angle. These include the widely used x-ray fluorescence (XRF) methods and those based on measurement of the intensity of a diffraction peak. These are not traceable but may be calibrated against known standards, and are useful when the interferometric methods are inapplicable. Parameters conveniently measured by these methods include:

- Composition of amorphous metal alloys
- Thickness of layers of known composition and density above the thickness possible by interference methods
- Thickness of rough polycrystalline layers

The assumption that is inescapable for thickness measurements by intensity methods is that the density of the sample is the same as (or in a known relation to) that of the standard. Additionally, for diffracted-intensity methods the crystalline state of the sample must be the same as that of the standard.

An important property of x-ray measurements is that the scattering properties are relatively insensitive to chemical effects, such as bonding state.

This is because they involve photon energies of thousands of electron volts (eV), whereas chemical effects are of the order of a few electron volts. Thus, to x-ray a silicon atom in pure Si looks almost the same as a silicon atom in SiO_2. However, the optical constants used in ellipsometry and optical reflectometry are very different, and also differ between bulk and thin-film states. While the tiny differences caused by the bonding state can be measured in certain refined x-ray experiments (x-ray absorption near-edge structure, or XANES), they cause no perceptible error in x-ray metrology (XRM). It is amply sufficient to take scattering factors from free atom databases, scaled with material density; density information is itself sometimes contained in the experimental data. This is in sharp contrast to optical metrology, where the similarity of the probing photon energy and the chemical effects in the materials, and the large variations of refractive index with material state are major problems to be solved in the modeling. It is worth noting, however, that the x-ray energies and intensities are far below the thresholds needed to damage the electronic properties of semiconductor wafers.

Moreover, the scattering parameter values themselves are "Goldilocks"* values — not too little, not too large, but just right. The scattering and interference phenomena are, as we shall see, easily measurable in most cases of interest to XRM and provide sensitivity to important parameters right down to the smallest scales presently used. Examples include thin high-*k* dielectrics down to 1 nm, pore sizes of low-*k* dielectrics down to 0.5 nm, SiGe composition to 0.1% of value, strain in strained silicon down to tens of parts per million, and lattice parameter down to a value in which the variation can still be seen in the best crystals yet grown. As thin films approach thicknesses of a few atomic diameters, so it becomes more appropriate to measure their properties with a probing radiation whose wavelength is itself of atomic dimensions. Measuring a current gate oxide with an optical probe is like measuring the width of a pencil line with a meter ruler.

The analysis of x-ray data to give metrological information rests upon the sound knowledge of the scattering parameters and the excellent predictive theory that has been developed for x-ray scattering over the last century. It has been possible for a couple of decades to simulate the x-ray scattering very accurately given a realistic material model. More recently, fully automated analysis of x-ray data has become possible, as discussed in Chapter 13. The inset on the next page shows the steps in this approach.

Precision, or repeatability, is a major concern for fabrication engineers, and here again XRM performs very well. It has the further advantage that measurements based upon interference fringes are inherently absolute and traceable, a topic we discuss in detail in the final chapter. These measurements depend only on wavelength and angle, which are easily referred to absolute standards. With the important proviso that proper procedures must consistently be used in alignment, x-ray tools will automatically give accurate results without the need for standards, golden specimens, or golden tools.

---

* An apt description, due to the late Richard Deslattes of NIST.

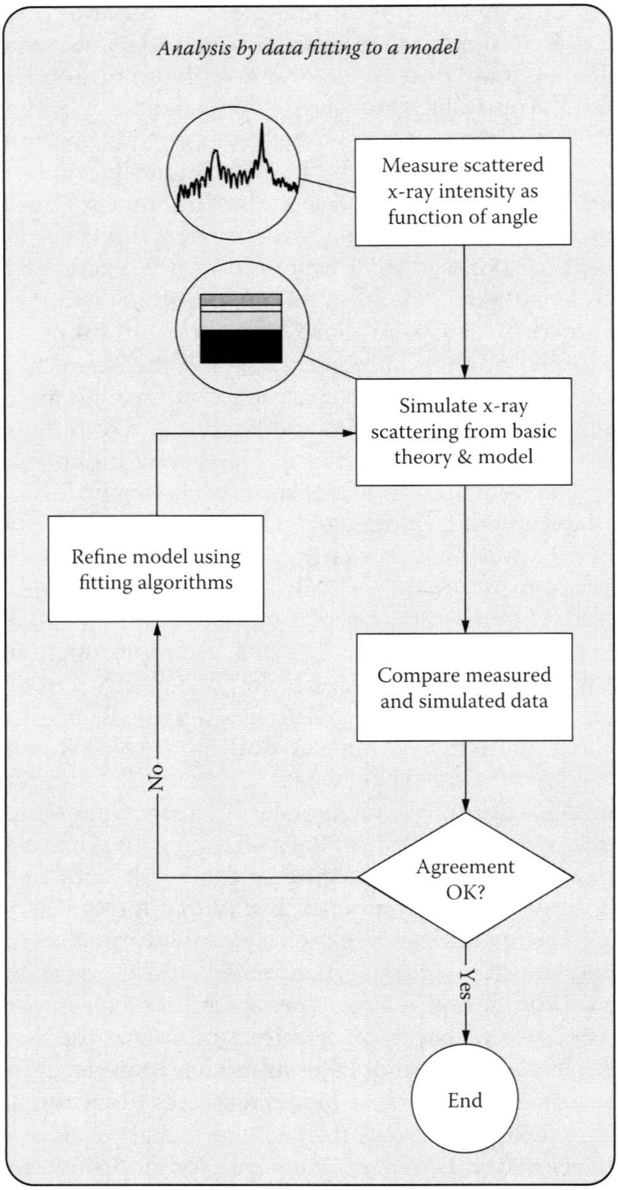

Semiconductor manufacturers are increasingly aware of the desirability for absolutely *accurate*, not merely *repeatable* or *precise*, metrologies.

X-ray sources are considerably less bright than optical sources (*brightness* is the number of photons per second, per unit area, per unit solid angle). Throughput and spot size have therefore been the main limitation of XRM. In recent years, tool manufacturers have put major efforts into the development of brighter sources and their more efficient utilization by means of novel x-ray optics, which improve both these parameters. Only 10 years ago, XRM could not be considered practical because of its low throughput and large spot size. At the time of writing it has been accepted and installed in many 300-mm production fabs as giving adequate throughput with <100-μm spot size. Such specifications now approach those of optical methods. Moreover, the field is still developing rapidly, and road maps exist to improve the throughput substantially over the next few years. However, engineers should always remember that, due to fundamental limitations set by the physics of x-ray generation in a metallic target, there is a fixed number of photons that can be utilized from the source, and that of the three properties of interest — repeatability, spot size, and data acquisition time — it is possible for a given system to optimize only two. The third parameter will be determined by the source brightness.

In the remainder of this chapter we shall give a brief introduction to XRM, classified by technique. In the rest of Part 1, the metrologies are classified by application rather than by technique, so that a fab engineer can easily assess their usefulness for a particular metrological problem.

1.2 Specular X-ray Reflectivity (XRR)

Film thickness has always been the most important parameter in semiconductor metrology, and can be very easily measured by x-ray reflectivity down to subnanometer thicknesses. The inset boxes overleaf show the typical measurement method and parameters.* X-rays incident at a grazing angle are specularly reflected from the surface and interfaces in the material. The mirror reflection is near 100% at low angles, since x-rays experience total *external* reflection from solids or liquids, with a critical angle for total reflection of less than a degree.** Waves partially reflected from different interfaces interfere, giving the Kiessig fringes seen in Figure 1.2. The physics is the same as that observed in thin films of oil on the surface of a wet road, where colored fringes are seen in (white) sunlight. The fringes are readily seen on this example, as is the accuracy of modeling. Essential features are:

- As the layer becomes thinner, the period of the fringes increases.
- The practical range of measurement is 1 nm to 1 μm for most materials.
- Specular reflectivity (XRR) measures electron density changes normal to the surface.

* A detailed explanation of the technologies, procedures, and experimental errors for each technique is given in Part 3 of the book.

** In Part 1 of the book we simply assert the scientific results without explanation. The detailed explanation and theory, together with the references and bibliography, are given in Part 2.

- XRR is not sensitive to structural parameters; single crystalline, polycrystalline, and amorphous thin films can be measured equally well.

Because the refractive index of all materials for x-rays is equal to unity within a few tens of parts per million, the spacing of Kiessig fringes is almost constant (except at very low angles) and can be used to estimate the film thickness with no knowledge of the material whatsoever — a true Ångstrom ruler. The fringes are a little more closely spaced at the very low angles, due to the small effect of refractive index. A simple (geometrical) correction for

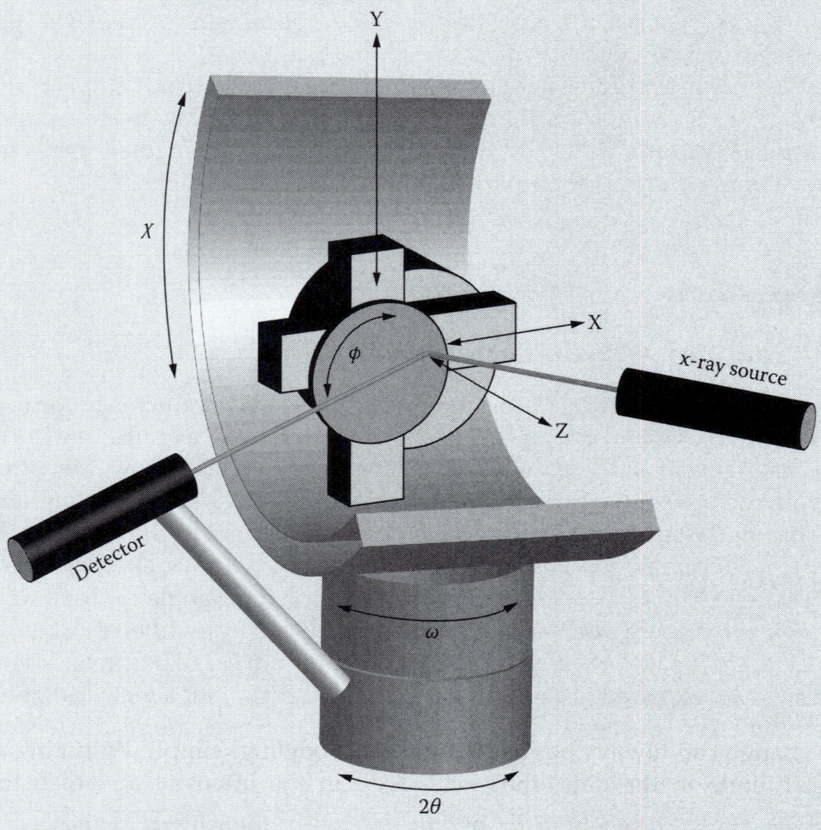

X-ray measurements require an x-ray source, a beam conditioner to control wavelength and divergence of the input beam, a goniometer to manipulate the sample, a detector to measure the scattered intensity, and often a detector collimator to limit the divergence of the output beam that is measured.

Refer to this box when reading the data collection description under each of the techniques.

XRR: Measurement of specular and diffuse scatter

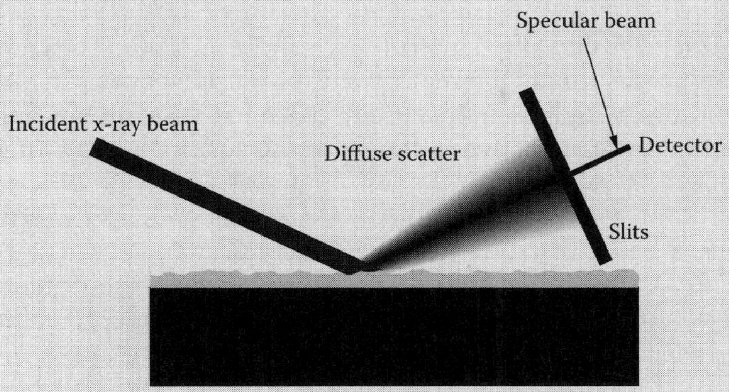

The incident and scattered angles are exaggerated for clarity; they are around 1°. The incident and scattered beams are collimated with slits or crystals to ~25 arc sec.

Specular scatter is measured with equal incident and scattered angles, in an $\omega - 2\theta$ scan.

Diffuse scatter is any scatter not in the specular direction.
It may be measured in several ways

- Scanning of the ω (specimen) axis – "transverse scan" or rocking curve
- Scanning of the 2θ (detector) axis – "radial scan"
- Coupled scan of $\omega - 2\theta$ with offset of omega from specular condition – "longitudinal scan"

Resolution settings are similar to those for specular scans, though for special purposes a lower exit divergence is used.

this change can always be made, without modeling, simply by noting the critical angle in the data, the sharp drop in the intensity at which total external reflection ceases. This gives both an extremely accurate thickness value and the refractive index. Hence, we obtain the electron density of the material, which is easily converted into physical density if the chemical composition is known.

The width of the interface at the top surface of the material is given by the slope of the envelope of the curve. For a perfectly sharp interface of a single material such as a wafer substrate, the decrement of intensity above the critical angle is proportional to the inverse fourth power of the scattering angle. In practice, this gives measurement of surface roughness between approximately 0.01 and 4 nm. If a single layer is present, the amplitude (peak to trough) of the Kiessig fringes is determined by the electron density differences between the substrate and layer, modified by the effects of the roughness and grading of the interfaces. The fringe amplitude generally increases with electron density differences but may increase or decrease with scattering angle.

The interface width is often interpreted simply as topographic roughness, and in many cases this is the major contributor. However, composition or density grading at the interface, caused, for example, by interdiffusion at an interface or by leaching of a component at the surface, has an effect on the *specular* reflectivity identical to that of roughness. If topographic roughness and chemical intermixing must be separately measured, diffuse scatter (Section 1.3) must be used in addition to specular scatter. A clue that this distinction is important is given if XRR roughness values are significantly higher than those found by atomic force microscopy (AFM) for a given material.

Thus, simple interpretation of the x-ray reflectivity profile as a function of angle easily gives thickness, interface width (roughness or intermixing), and density.

If multiple layers are present, the amplitude of scatter from each is added in the output signal. This produces the phenomenon of beating of the waves, and very complex fringe patterns may ensue. Simple cases may be sorted out by transform methods, but more generally modeling methods are used to analyze the data. The complexity of the signal is not a problem but an asset: the richness of the signal allows complex multilayers to be measured in detail. The thickness and interface width of buried layers can be determined by XRR, until the layers are so thick that the reflected signal from the lower interfaces is lost through absorption. This occurs in the range of 0.5 to 2 μm, depending upon the absorption of the material. To a good approximation absorption is linear with the average atomic number, and inversely proportional to the cube of the wavelength. Examples of the penetration of a grazing incidence beam are shown quantitatively in Figure 1.1.

For a multilayer, defined as a set of repeats of a layer structure in a superlattice, strong reinforcement of the scatter occurs periodically, at angles given by $n\lambda = 2t \sin\theta$, where t is the repeat period, λ is the wavelength, and n is an integer. These are known as Bragg peaks, by analogy with scattering in crystals, since the layers in the superlattice behave as an artificial crystal. An example is shown in Figure 1.2. It is simple to determine the superlattice periodicity directly from the spacing of the superlattice Bragg peaks. The mark/space ratio of the basic period is given by the relative intensities of the superlattice peaks, and their damping or broadening with increasing angle (order of diffraction) is determined by the degree of constancy of the period.

The ability to measure the density of the surface region of the material has important consequences, such as the measurement of porosity if the matrix density is known. For a typical layer of low-*k* dielectric the beam is first reflected from the layer, then penetrates to the substrate at a higher angle (~0.22° for silicon). This is shown in Figure 1.3. Note that the Kiessig fringes

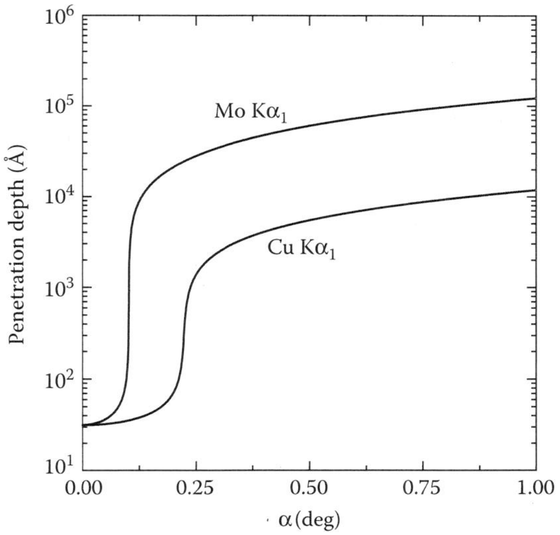

FIGURE 1.1
Penetration of an x-ray beam at grazing incidence. The graph shows the depth at which the amplitude of the electric field of the beam is reduced to 1/e of its value at the surface, for two wavelengths (CuKα at 0.154 nm and MoKα at 0.071 nm). The graph is shown for silicon. Materials of higher electron density (approximately proportional to average atomic number) will have the sharp rise at the critical angle shifted to higher angles.

begin immediately after the first critical angle (just over 0.125°). For stoichiometric compounds such as SiO_2 the bulk density is well known and the porosity follows accurately and traceably. For carbonaceous low-*k* dielectrics, the bulk density is less certain, so the absolute measurement of porosity involves an assumption, or some external data.

1.3 Diffuse Scatter

When a layer is grown, its peaks and valleys may or may not conform to those of the layer beneath. This may be advantageous or detrimental to the function of the layer; for example, for gate oxides conformal roughness is desirable, as the layer should not have exceptionally thin spots. Peaks and valleys seen in the superlattice reflectivity curve shown in Figure 1.2 will also be seen in the diffuse scatter if the layers are conformal. This can be represented mathematically by a vertical correlation function, and the associated vertical correlation length, ζ, can be measured from the diffuse scatter. Both specular and diffuse scatter data from another superlattice are shown in Figure 1.4. The superlattice peaks are seen in both. By suitable modeling, the vertical layer–layer correlation function can be found. This

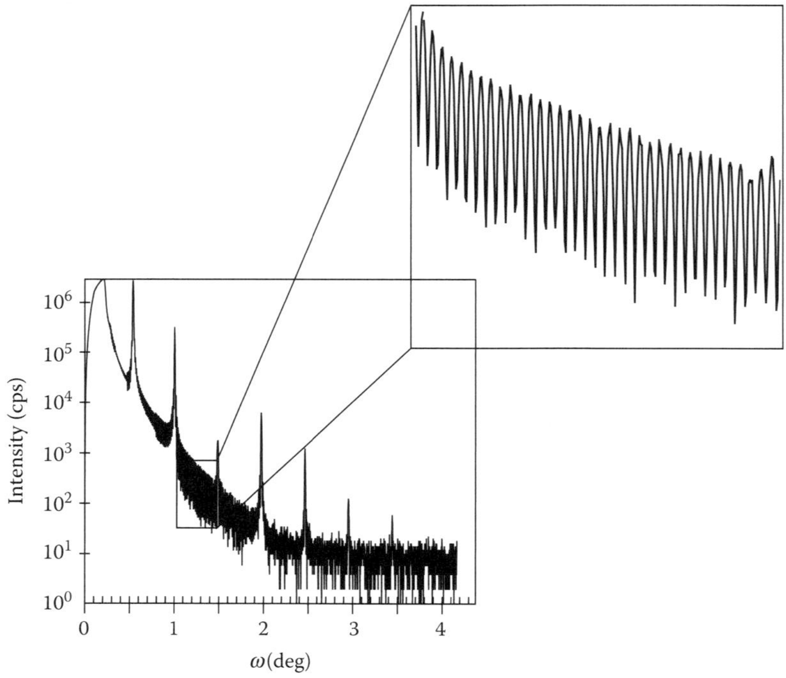

FIGURE 1.2
XRR from a 40-period (60 Å Si + 30 Å $Si_{0.7}Ge_{0.3}$) superlattice.

is computationally intensive and is appropriate for process development rather than online quality control.

We have seen how specular x-ray scatter gives a good measure of interface width, but that we cannot distinguish between topological roughness and the grading of composition or density at the interface. As shown in Figure 1.5, both roughness and grading smooth the refractive index profile normal to the interface when averaged over the area illuminated by the x-ray spot. Diffuse scatter allows us to distinguish between these two. Grading normal to the interface cannot scatter x-rays outside the specular direction. We can understand this by imagining a graded interface as a set of mirrors of differing reflectivity that are all parallel to the ideal macroscopic surface. Roughness, however, will contribute to diffuse scatter, since this introduces mirror regions that are not parallel to the macroscopic surface and reflect at different angles. Analysis of such scatter allows the distinction between the two effects (Chapter 9). This is more likely to be used in process development rather than online metrology since the data collection and analysis times are relatively long.

In addition to the Gaussian root mean squared (rms) roughness, simulation of the diffuse scatter requires a model for the height–height correlation function of the interface. This function answers, mathematically, the following

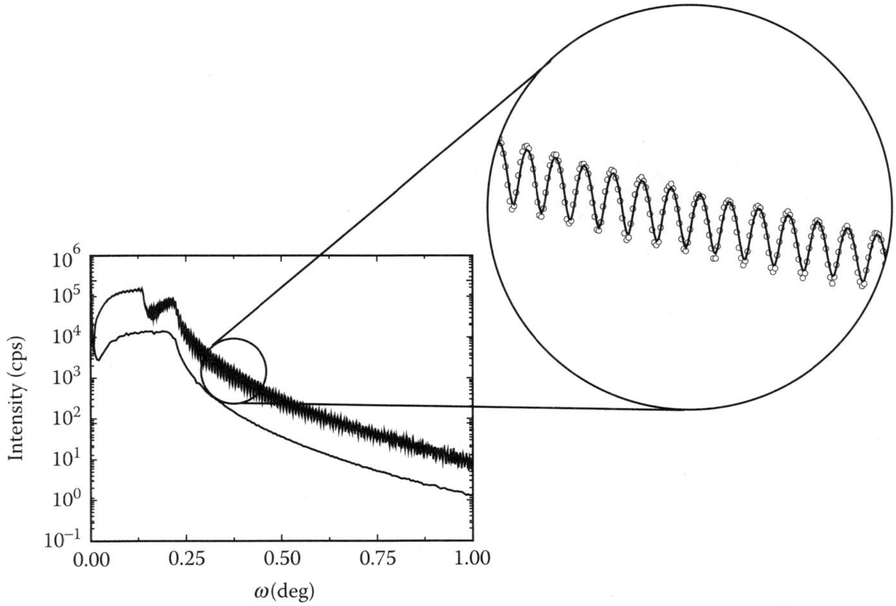

FIGURE 1.3
Specular x-ray reflectivity data from a porous, low-*k* dielectric thin-film (upper line) sample. The data from a Si substrate is shown for comparison (lower line). The inset shows the agreement between experiment (points) and best-fit simulation (line) on an expanded scale.

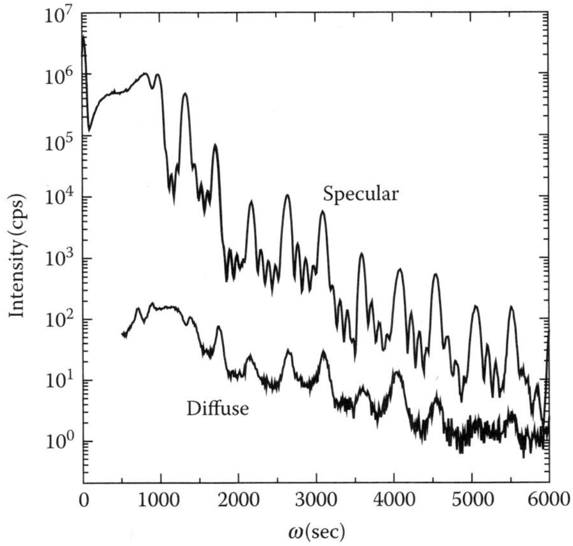

FIGURE 1.4
The specular and longitudinal diffuse scan of a conformal Si/$Si_{0.43}Ge_{0.57}$ five-period superlattice.

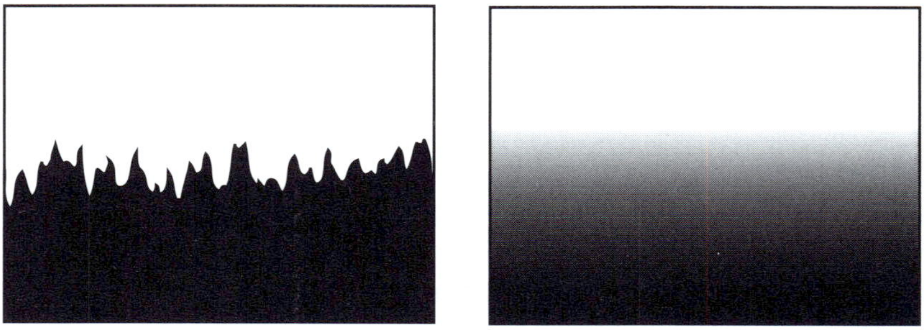

(a) (b)

FIGURE 1.5
Roughness (a) and grading (b) reduce the specular reflectivity, since both smooth the refractive index profile normal to the interface. Roughness generates diffuse scatter, grading does not, but it damps the scatter caused by roughness.

question: If you know the height at a certain point on the interface, how well can you predict the height at another point? The most common model uses a self-affine fractal structure in which there are two key parameters:

1. The in-plane correlation length, ξ, a measure of the scale length of the structure within the surface
2. The fractal Hurst parameter, h, representing the jaggedness of the surface and related to the fractal dimension, D, of the interface by $D = 3 - h$

Both ξ and h can be extracted by modeling of the diffuse scatter. They are determined by the film growth conditions and may be relevant to the nature of film growth on the next layer.

It has long been known that small particles or pores in a material cause small-angle scatter, around both the incident beam and the diffracted beams. Both can be used, in principle, to determine pore size and its distribution, though the diffracted beam is only usable when the material is a single crystal, e.g., porous silicon. Transmission x-ray (SAXS) and neutron (SANS) small-angle scattering have been the traditional approach. More recently, the measurements have been performed in reflection (where it has become known as grazing incidence small-angle x-ray scattering, or GISAXS). This is considerably more convenient for fab tools, and allows greater scattering volumes, but has the disadvantage that more complex modeling is required because of the effects of total reflection. It also has the restriction that the smallest scattering angle is equal to the incidence angle, which should be above the critical angle of the film in order to obtain sufficient scattering volume. There is thus an upper limit of 15 to 20 nm on the pore size that can be measured.

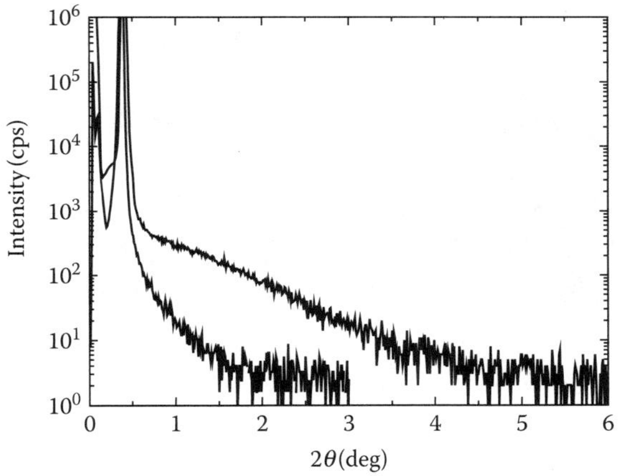

FIGURE 1.6
Diffuse x-ray reflectivity data (2θ scan) from a porous low-k dielectric thin film (upper line). The broad intensity distribution is characteristic of scattering from micropores in the thin film. The data from a Si substrate is also shown for comparison (lower line). Both scans were performed with a fixed incidence angle, $\omega = 0.19°$.

Any of the diffuse scatter scans will register the effects of porosity, but our preference is for the radial or detector-only scan. The incidence angle is fixed; therefore, the penetration is constant (and can potentially be varied to give depth-sensitive information). An angle in between the critical angles of film and substrate is chosen, so that the electric field is intense in the film. Typical data are shown in Figure 1.6.

The measurement is strongly dependent upon pore size, and trials show that the repeatability is as good as ±0.1 nm. Determination of the absolute value depends on some assumptions. First, the surface roughness, which also causes diffuse scatter, must be either very low or known. Second, the diffuse scatter can only be calculated if the pore size distribution function is assumed. Figure 1.7 illustrates two common types of distribution, which for the same average pore size give somewhat different shapes to the diffuse scatter curve. These may be distinguished, at least in some cases, by seeing which distribution gives the better fit.

The measurement of pore size is not yet fully developed as an absolute metrology. For a consistent pore size distribution, and in the absence of complicating factors such as surface roughness, variation of pore size with depth, and complex distributions, the measurements are sufficiently sensitive and repeatable to be used as a relative and repeatable metrological method. Further research is still needed to find methods of quantifying some of the complicating factors. This technique is currently being used for materials and process development, and is being developed as a potential in-fab metrology for porous low-k ILDs.

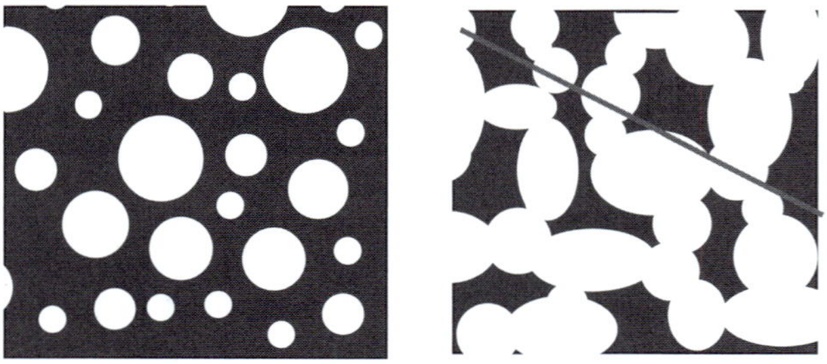

FIGURE 1.7
(a) Nonconnected pore structure. Spherical pores with average pore diameter, $\langle D \rangle$. Polydispersity, d, measures the width of the pore size distribution. (b) Connected pore structure. Correlation length, x, is a measure of the average spacing between regions of phase 1 and phase 2. Average pore dimension, x/P, and average wall dimension, $x/(1 - P)$.

1.4 X-ray Diffraction

Polycrystalline thin films are widely used in semiconductors, as conducting layers (Al, Cu, and W), diffusion barrier layers (TaN/Ta, TiN/Ti), and some semiconducting and dielectric layers. X-ray diffraction (XRD), in which strong scattered intensity is observed at specific angles of scattering, has been widely used for decades in the determination of structural and microstructural parameters in powders and polycrystalline bulk materials. Recently, it has been realized that many parameters in the thin films used in semiconductor manufacture can be measured by XRD, even in films <2 nm thick. These include:

- Measurement of the proportion of crystallinity in films that are nominally amorphous
- Determination of the composition of the crystalline components of the material
- Measurement of the grain size of a crystalline material
- Determination of the stress in a crystalline material
- Measurement of crystallographic texture, that is, the distribution of crystallite orientations within a polycrystalline material

The interpretation of the XRD is based on the Bragg law (see inset), which determines the scattering angles at which the peaks of strong scattered intensity may occur. Composition is measured through identification of the positions and intensities of diffraction peaks, which are unique to a given chemical compound. Automated comparison with a large reference database

The Bragg law

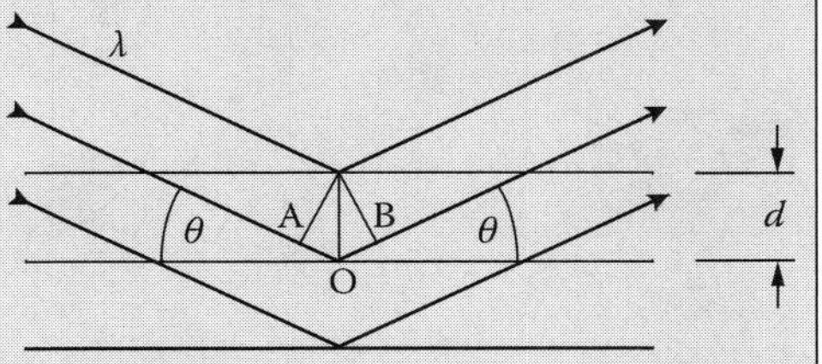

Strong diffraction from a set of planes results when

- The angles of incidence and diffraction, θ, are equal
- The path difference AOB between the two beams is equal to an integral number of wavelengths, $n\lambda$

Hence the Bragg law

$$n\lambda = 2d \sin \theta$$

of known compounds (the search–match process) is used for analysis. It is a structural identification, which can answer targeted questions, such as:

- Is deposited Ta in the high-conductivity β-phase?
- Is the tantalum nitride barrier layer of the right composition?
- What is the likely identification of this unknown contaminant coating?

Once a phase mixture is identified, quantitative analysis may be performed to measure the compositions, using the intensities of the peaks and information about the texture if necessary. If the material is supposed to be amorphous, it is also possible to see if there are any diffraction peaks at all, which indicate microcrystalline regions.

The smaller the size of a set of diffracting particles, the broader are its diffraction peaks. These may be analyzed quantitatively to measure grain size. Stress will have two effects. A microstress or an internal stress within the grains that averages to zero over the diffracting volume will have no net effect on peak position but will broaden the peaks. The effect of both grain size and stress is to broaden the Bragg peaks. However, the broadening is a different function of scattering angle, 2θ, for the two effects, and hence they may be separately measured. The Williamson–Hall plot, in which the peak full-width-at-half-height maximum $\Delta(2\theta)$ multiplied by $\cos\theta$ is plotted

Measurement of XRD from polycrystalline films

I Bragg-Brentano method

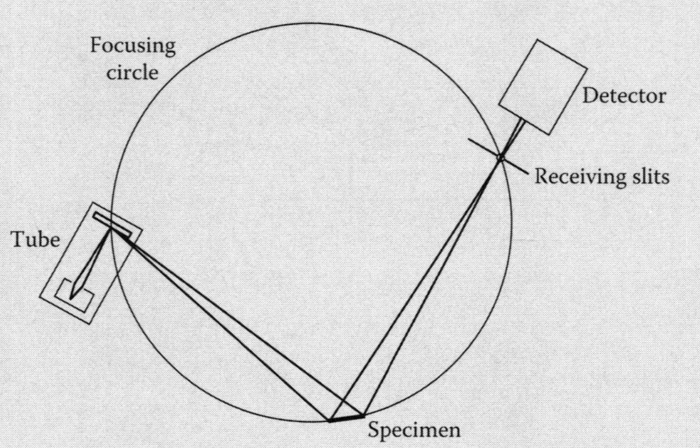

This is the traditional method. The specimen is scanned from zero to approximately 50°, while the detector is scanned at twice the rate from 0 to 100° – an "Omega-2theta" scan. The specimen is symmetrically placed between source and detector. Thus, crystal planes that (a) satisfy the Bragg angle, and (b) lie on the tangent of the focusing circle will diffract into the detector.

This method gives geometrical focusing, since angles standing on the same arc of a circle are identical. Hence, the diffraction efficiency is high and intensities good. However, the method suffers from a number of aberrations and strongly textured specimens may give only one peak. It is not very suitable for thin films since the penetration depth is large (tens of μm) and varies with the incidence angle.

against $\sin\theta$, is used to quantify the grain size and strain dispersion, as shown in Figure 1.8.

An externally applied stress, or an internal stress that does not average to zero over the diffracting volume, will give a peak displacement due to the strain induced. This may be measured and converted to stress if the (local) Young's modulus is known. Normally, only the biaxial (in-plane) stress is measured. The peak shift is measured as a function of the angle of the diffracting planes relative to the surface normal (ψ) and plotted against $\sin^2\psi$. The slopes of two such curves, taken parallel to two principal axes of stress (see Chapters 4 and 10), determine the two biaxial stresses. An example is shown in Figure 1.9 for a Mo film on glass.

Finally, it may be necessary to determine the crystallographic texture of a polycrystalline material. This is the extent to which the orientations of the crystal axes of the individual grains are not random but point in one or more

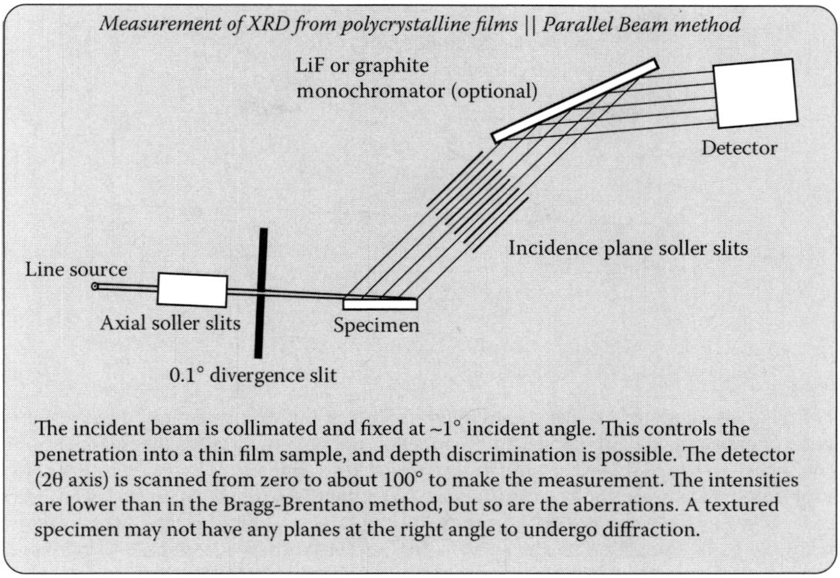

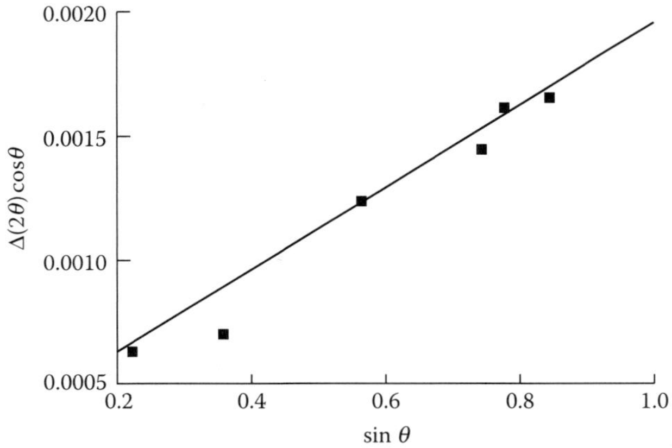

FIGURE 1.8
Williamson–Hall plot for a 5% SiC-alumina nanocomposite. The intercept on the y axis gives the mosaic domain size, here 1 ± 0.5 μm, and the slope the strain dispersion, here $9 \pm 1 \times 10^{-4}$. These are typical error numbers for this type of analysis.

preferred directions, relative to the physical sample. For example, sputtered metals with the face-centered cubic structure usually have a strong tendency for the close-packed (111) planes to lie parallel to the surface, because these are the slow-growing planes. Electrodeposited copper usually shows a similar but weaker texture. Texture of a layer may be important for the physical properties, such as magnetism or elasticity. It is also important for process

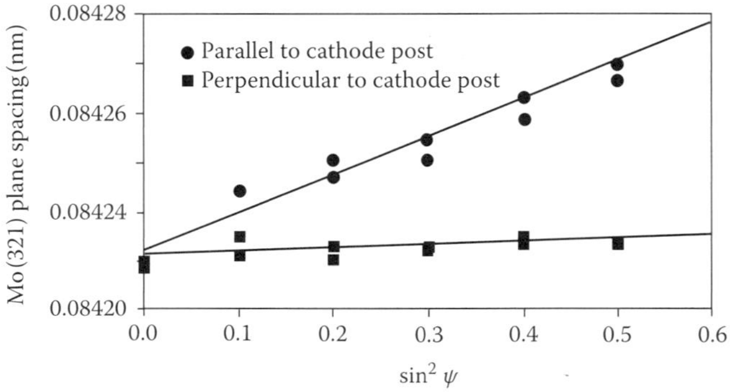

FIGURE 1.9
Plot of (321) lattice plane spacing in Mo (measured from the shift of the diffraction peak) as a function of exit angle (ψ), plotted using the function $\sin^2 \psi$ The sample is magnetron-sputtered Mo on glass. The biaxial surface stresses are found from the slopes of the two lines. In this example the stress is anisotropic, with different values parallel (481 ± 56 MPa) and perpendicular (99 ± 56 MPa) to the cathode post. (From Ballard, B.L. et al., *Advances in X-Ray Analysis*, 37, Gilfrich, J.V. et al., Eds., Plenum Press, New York, 1994. With kind permission from Springer Science and Business Media.)

behavior such as response to etching, or growth nucleation for the next layer, and may influence the elecromigration speeds and hence service lifetimes of devices.

An example is shown in Figure 1.10. Texture is measured by setting the detector on the peak whose distribution is to be measured, e.g., (111), and systematically rotating the specimen through all available incident angles, measuring the intensity of the peak. Results are plotted in a pole figure, as in Figure 1.10. For a complete description of texture, it is necessary to take three independent pole figures and then calculate the orientation distribution function (ODF), but this is time-consuming and rarely appropriate in XRM.

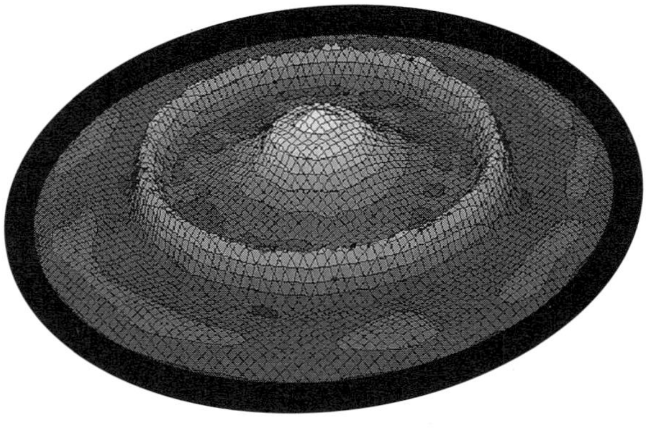

FIGURE 1.10
Pole figure of damascene copper, showing the density of distribution of the (220) planes as a function of angle relative to the wafer. The central point is normal to the wafer surface.

Rather, a simplified single scan may be taken, representing a section of the pole figure, which can give enough information in a simple quantifiable form. This is particularly useful for fiber textured samples, which is the most prevalent texture in polycrystalline thin films.

X-ray diffraction may also be used to measure the thickness of films on the micrometer scale, above the range of thickness possible by x-ray reflectivity. One method is to measure the integrated intensity under the Bragg peaks, using the intensity under the Bragg peak from the single crystal silicon substrate as a calibration standard. This is convenient and simple to use, but does rely on the texture of the film remaining constant and is of course not an absolute metrological method. An absolute method of thick-film metrology involves the determination of the reduction of the integrated diffracted intensity from a substrate Bragg peak by the presence of the film. Provided that the film composition and density are known, the absorption gives the thickness directly. The technique is most appropriate for ranges of thickness and atomic number of the film material, where the presence of the film reduces the intensity in the substrate Bragg peak by a factor up to 10. In copper and tungsten this corresponds to maximum thicknesses of approximately 12.5 and 2.5 μm, respectively, for the 004 substrate reflection with CuKα radiation.

1.5 High-Resolution X-ray Diffraction

High-resolution x-ray diffraction (HRXRD) has been used since about 1980 in the metrology of compound semiconductors, for which it is vital to the control of composition and thickness of ternary and quaternary layers. Selected-area epitaxy of SiGe was introduced in the 1990s, for two purposes. These were to improve high-frequency transistor performance through use of heterojunction bipolar transistors (HBTs) and to act as a virtual substrate to grow thin strained layers of silicon. In these materials, Ge composition and thickness can be measured to a repeatability and accuracy well within process tolerance values by HRXRD. Figure 1.11 shows a measurement on a box structure of SiGe epitaxially grown on a Si wafer. From the composition of the SiGe, the strain in the cap layer of silicon can be found and the resulting enhanced electron mobility may be calculated.

The need for high *angular* resolution in HRXRD arises because the peak widths, separations, and details examined are at the arc second scale, as seen in Figure 1.11. The substrate peak is used as an angular reference. The difference between this and the layer peak gives the Ge concentration, through the known calibration of the effect of Ge on the expansion of the lattice parameter. Information on the layer thickness is found in the intensity of the layer peak and, much more sensitively and accurately, in the period of the interference fringes. Multiple layers, including graded layers, give

Measurement of HRXRD from epitaxial films

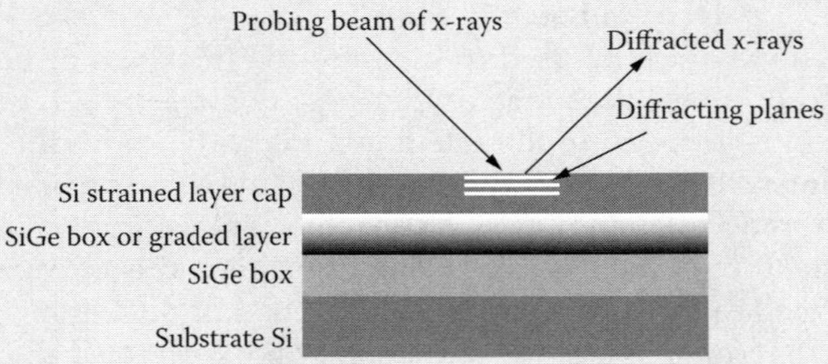

The incident monochromatic beam is highly collimated, to ~12 arc sec to give enough angular resolution for the measurement. The detector beam is collimated to ~0.05° to improve signal/noise. Data are collected with an omega/2theta scan over typically 1°–2° range on omega. Several decades of intensity are collected.

The above "symmetrical" setting measures strain perpendicular to the surface. The "asymmetrical" setting below measures a component parallel to the surface. Combination of both allows epitaxial relaxation to be measured.

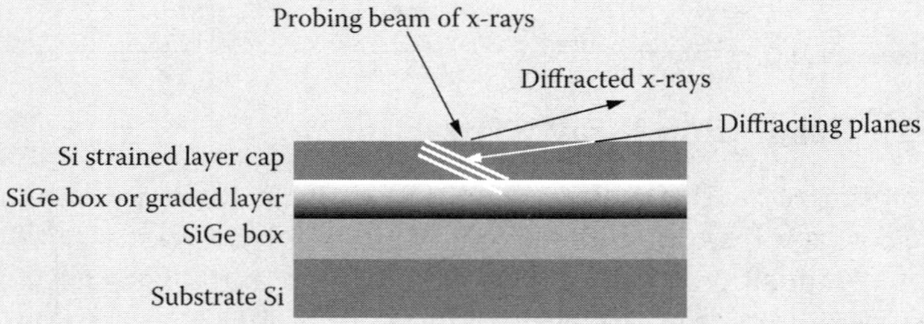

more complex interference patterns that can be accurately modeled. Within limits the layers can be measured individually.

The use of modern x-ray optics allows the measurement to be made on small spots with sufficient throughput. The current practical limit for HRXRD in the laboratory or production fab is a spot approximately 30 × 30 μm. Standard optical image recognition and alignment techniques can be used to measure a specified characterization spot on the wafer, and this spot is small enough to be located in the scribe lines ("streets") in between dies. Thus, measurement can be made on patterned product wafers as well as blanket "witness" wafers. This is particularly important for silicon-based technologies, in which the SiGe epilayer is often grown not as a blanket across the wafer (unlike compound semiconductor epilayers) but in small selected areas. Edge effects in these growth windows usually mean that

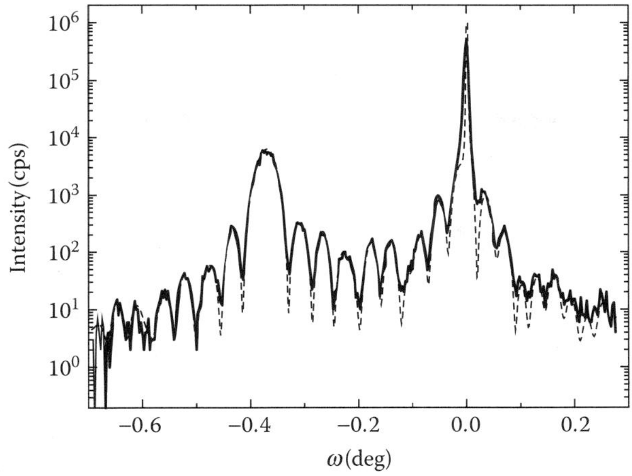

FIGURE 1.11
HRXRD scan of a SiGe box structure: thickness, 125 nm; composition, 14.2% Ge. Both experimental points and the modeled curve are shown. The large peak on the right is from the substrate; the smaller peak on the left is from the layer. Interference fringes from the layer thickness are clearly seen.

growth is somewhat different from that on a blanket wafer in the same reactor.

With good encoded goniometers, HRXRD can also measure absolute scattering angle. From the Bragg law, this can then be used to measure lattice parameter, with direct traceability to the international meter, to a precision better than a part per million. This can be used to quantify dopant levels and map the variation of lattice parameter across the wafer, as well as provide the calibration method for the effect of Ge and other alloying/dopant elements on the lattice parameter of silicon.

As device layers become thinner, the need for accurate wafer orientation measurement increases, for the alignment of crystal channels for ion implantation. The encoded x-ray diffractometer can achieve better than 0.01° accuracy and repeatability in this measurement.

Thin layers of strained silicon are now being used to make devices. In these, an artificial or virtual substrate is formed by growing relaxed layers of silicon-germanium with Ge content from 20%, up to 50%. A high strain is then induced in epitaxially deposited silicon as long as it is kept below the critical thickness for onset of relaxation. Layers a few tens of nanometers thick can be measured even in the presence of the large silicon substrate peak, but the peaks from thin layers are broad and are swamped by the substrate peaks. However, the grazing incidence in-plane diffraction can be used for layers as thin as 5 nm, since the beam penetration below the critical angle for external reflection is only a few nanometers. The technique is shown in the inset box.

For the virtual substrates used for strained silicon it is important that the layer be fully relaxed. However, in selected-area epitaxy to grow transistor

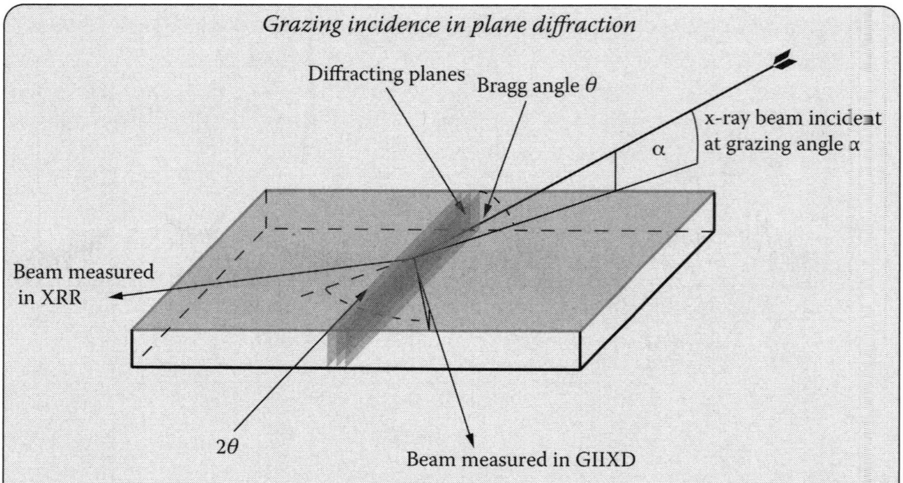

The incident beam is collimated and fixed at an incident angle close to the critical angle for external reflection. This controls the penetration into a thin film sample, and depth discrimination is possible. The sample is rotated about its normal (Φ axis) and the detector (2θ axis) is scanned from zero to about 100° to make the measurement – a $\Phi - 2\theta$ scan. The intensities are low but the method uniquely gives substantial information about thin textured films such as sputtered metals. Peak widths are considerably narrower than for non-grazing methods since many more planes contribute to the diffraction.

elements, relaxation gives rise to crystalline defects that can reduce yield. The SiGe layers used for these devices are usually in the metastable region (in between the critical thicknesses calculated for full equilibrium and kinetic equilibrium). Even if they are grown unrelaxed, there is a danger that they may relax on later processing. The ability to measure relaxation rapidly on product wafers is therefore an important metrology. Figure 1.12 shows a recently developed relaxation scan method of performing this metrology. The underlying principles are given in Chapters 4 and 11.

1.6 Diffraction Imaging and Defect Mapping

The principle of x-ray topography (XRT) is that of mapping the diffracting power of a crystal, in either transmission or reflection, across its surface. Defects in the wafer or an epilayer show up as contrast in the image due to the different local scattering power in the vicinity of a defect. Film or plate-based methods of XRT have been in use for about 60 years and give excellent images. Extensive work on the theory of x-ray scattering in highly perfect crystals now means that images can be simulated accurately. However,

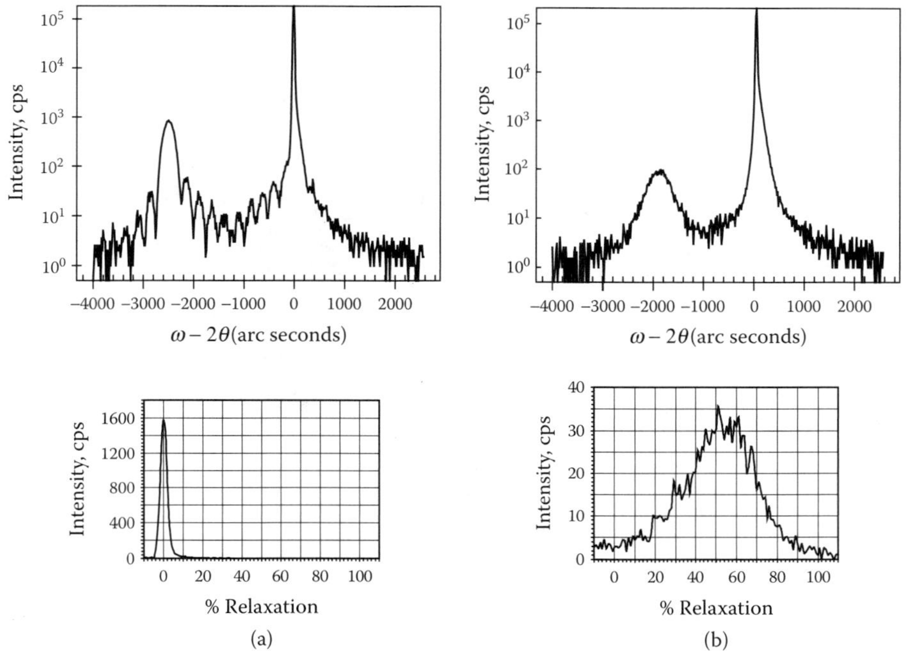

FIGURE 1.12
Measurement of relaxation of a SiGe layer. The upper figures are 004 reflections, used to determine the relaxation axis, a line in diffractometer space on which the relaxed peak must lie, and the lower figures are relaxation scans along this axis. The wafer has a $Si_{0.75}Ge_{0.25}$ layer. (a) Unrelaxed region. The thickness fringes and relaxation scan are sharp and show zero relaxation. (b) Relaxed region. The thickness fringes are smeared out and the relaxation scan is broad, showing a distribution of relaxation about an average of 55%.

film-based methods are unsuitable for use in fabrication lines; they are slow, film processing is required, and further digitization is needed before automatic interpretation can be made.

Recently, a new method of digital x-ray topography, called x-ray diffraction imaging, or inspection (XRDI), was introduced. The principle is shown in the inset. A digital imaging camera is used, and the size restriction of the latter is overcome by performing a virtual scan of the detector with integration and image reconstruction in the computer. The method is fast, compatible with fabrication lines, and gives a direct digital image that can be analyzed by standard image processing tools. An example is shown in Figure 1.13.

The following can be detected and measured by XRDI:

- Defects consisting of a region of crystal that is *misoriented* with respect to the perfect material
- Defects consisting of a region of crystal that is *strained* with respect to the perfect material

Digital X-ray Diffraction Imaging

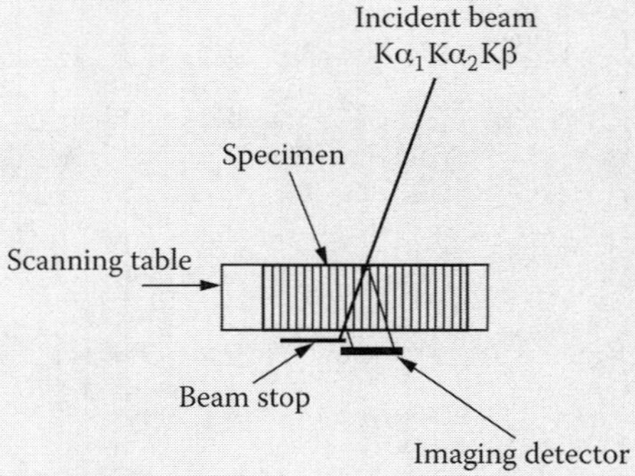

A microfocus x-ray tube is used to image a small stripe of the wafer on a static imaging detector. Multiple x-ray spectral lines can be used. The specimen is scanned and the image integrated and reconstructed in the computer. Transmission (above) is done with a hard radiation such as Mo Kα, and reflection (below) with a softer radiation such as Cu Kα.

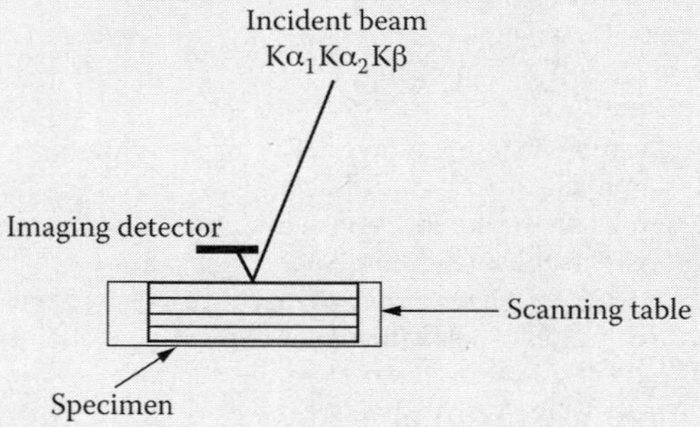

Local changes in diffracting power, caused by surface damage, dislocations, precipitates, etc. are seen as contrast in the image.

Spatial resolution is dependent on the detector, and can currently be as good as 3 μm. However, lower resolutions are normally used for whole-wafer surveys (macro defect inspection), since an image of a 300mm wafer mapped at this resolution contains about 60 terabytes of data. Once suspect areas are identified, they can be measured in detail with a local high resolution scan.

Examples include:

- Grown-in dislocation bundles
- Slip dislocations induced by thermal stresses
- Interface dislocation networks in heteroepitaxy
- Single dislocations from any cause

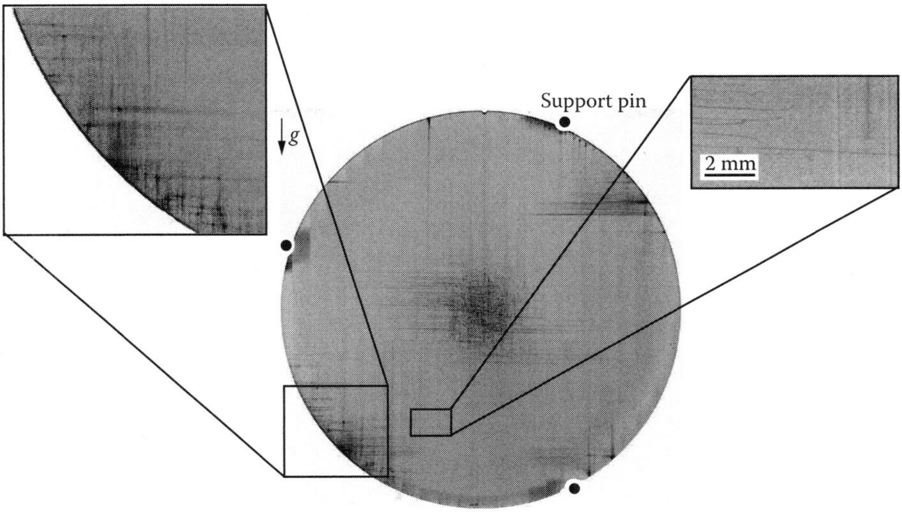

FIGURE 1.13
Example transmission x-ray topographic images of a whole wafer, with details of dislocation structure shown in selected regions.

- Dislocations in relaxed selected-area epitaxy windows as small as 200 × 200 μm
- Precipitates
- Surface damage, e.g., from grinding and polishing
- Edge damage, e.g., from profiling and polishing
- Orientation changes across the wafer, with mapping to 0.01° precision
- Subgrain boundaries

Transmission methods are used for defects in raw wafers, for example, dislocations and polishing or grinding damage, especially at edges. Reflection methods are best for measuring defects near the surface or in epilayers.

1.7 X-ray Fluorescence

When x-rays of sufficient energy strike a material, then its atoms fluoresce in the x-ray band. The fluorescent x-rays comprise a spectrum of lines whose energies (or equivalently, wavelengths) are unique to the element. These are known as *characteristic* x-rays, and their energies are determined by the atomic structure of the element (see inset). The intensity of the emitted radiation is proportional to the intensity of the irradiating x-ray beam and

X-ray Fluorescence

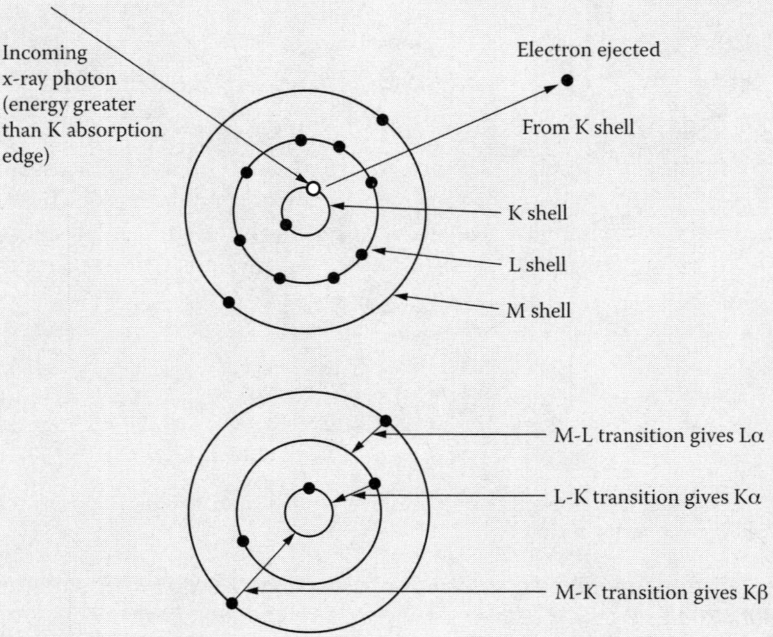

The incoming x-ray photon ejects an electron from an inner shell, leaving the atom in an excited state. Electrons cascade down from upper levels, emitting fluorescent x-rays (and also Auger electrons).

The energy differences between such levels are equal to the energies of the emitted fluorescent photons. Since the energy levels are unique to each element, the fluorescent spectrum provides a fingerprint of the atomic composition.

To eject photons from any shell, the incoming photon energy must be higher than the binding energy for that shell. These threshold energies are called "edges". For example, incoming photons of energy in between the K and L edges will generate L fluorescent lines but not K lines. Incoming photons of energy above the K edge will generate all fluorescent lines

The fine structure ($K\alpha_1$, $K\alpha_2$, etc.) is caused by the slightly different energies of electrons with different sublevels.

to the quantity of the element that is irradiated. This gives rise to the x-ray fluorescence (XRF) method of analysis. It may be used:

- For qualitative chemical analysis, in which the elements present are identified from their characteristic x-ray spectra

- For quantitative chemical analysis, in which the composition is obtained from the relative intensities of the characteristic lines from all of the elements present
- For film thickness analysis, in which the thickness is inferred from the intensity of a characteristic line of the element comprising the film

No calibration (other than of the energy scale of the detector) is required for the first of these, which is usually regarded as an analytical rather than metrological method. Where compositional analysis is required, XRF is very useful, and measurement of pure element standards suffices to calibrate factors such as air absorption and detector window absorption. The main limitation is the analysis of light elements, since the characteristic lines of such elements are at low energies and easily absorbed by air, by detector windows, and by the sample itself. XRF may be performed in an air environment for elements down to sodium ($Z = 11$), but for important elements such as carbon, nitrogen, or boron, a vacuum system is required.

The measurement of film thickness depends on calibration and an assumption. Calibration is performed by measurement of the fluorescent intensity from films of known thicknesses under identical conditions of incident beam, entry and take-off angle, detector type, aperture and distance from the sample, etc. A curve of intensity against thickness can be produced, which will be linear up to the thickness at which absorption of the incident or fluorescent rays in the sample becomes significant. The shape of this curve can be calculated from knowledge of some of the parameters, so a single calibration film can suffice.

The assumption is that the composition and density of the film are identical to those of the calibrating standard. The XRF method essentially measures the quantity of material in the x-ray beam, and the same signal can be produced from a thicker porous layer as from a thinner dense layer. The method can only be used in metrology when there are physical grounds for this assumption.

The main use of the XRF method of thickness metrology is on films that are too thick or too rough for XRR analysis, or where the electron density of the film is very similar to the neighboring films. An example of the former is damascene copper, in which thicknesses of >0 μm can be measured. An example of the latter is certain magnetic multilayers such as spin valve structures made of transition metal alloys. The combined use of XRR and XRF can usually provide a unique metrology of these materials.

1.8 Summary

- XRM is now established and in use in 200- and 300-mm fabs for metrology at the 90-nm node and below of many parameters that are important in wafers and thin films of epilayers, metals, and dielectrics.

- XRR gives reproducible and precise measurement of thickness, roughness, and density of surfaces and buried interfaces; overall porosity of low-*k* dielectrics may be inferred from densities.
- Diffuse x-ray scatter measures surface and interface roughness properties and pore size distributions.
- XRD identifies chemical compounds present in crystalline form and measures grain (particle) size, stress, and texture in such films.
- HRXRD measures composition, thickness, and relaxation of epitaxial layers.
- XRDI measures crystal defects (e.g., slip and other dislocations, polishing damage) in bulk wafers and epitaxial films.
- XRF measures composition of any type of film for elements of $Z \geq 11$, and the thickness of films too thick for XRR metrology (1 to 10 μm).

2

Thickness Metrology

2.1 Introduction

This is the most common requirement in semiconductor fabrication lines. X-ray reflectivity (XRR) is the most versatile x-ray method for thickness measurement, since it does not depend upon the crystal structure or state of aggregation of the material, only upon the electron density variation with depth. As we have seen, we may use XRR to measure thicknesses of almost any films below 500 to 1000 nm in thickness. But there are always limits, either from the nature of the material or from the requirements of factory metrology, e.g., requiring a very small spot. In this (and the remaining chapters in Part 1) we address the questions:

- Can we use x-rays to measure this parameter in our wafers?
- What are the limits of measurement?

There is one general principle that applies to all radiation-based metrologies but is more acute in x-rays than in optical methods. This is that the supply of photons from x-ray sources is low compared with optical sources. While synchrotron sources are very bright, there are obvious practical limitations in the construction of a fab metrology tool with reasonably compact sources. Fundamentally, brightness is constant through any optical system (Liouville theorem). A simple way of looking at the problem is that there are three parameters of great interest to fab engineers:

- Small spot size
- High throughput
- Good repeatability

These parameters are coupled. It is possible to optimize any two of them, but then the third will suffer. We may formulate three general principles:

1. Repeatability improves linearly with the square root of the data collection time.
2. Repeatability improves linearly with the diameter of spot measured.
3. Throughput = data collection time + analysis time + wafer cycle time.

Therefore, a modest relaxation of throughput or spot size can mean a large improvement in repeatability, and a modest improvement in repeatability or spot size can have a significant penalty on throughput.

In several examples in this chapter and in many cases throughout this book, we use calculated or modeled data. As will be seen in Chapters 8, 11, and 13, these calculations are very accurate and are good enough to be used to calculate tool acceptance and warranty criteria on previously unmeasured materials. Unless otherwise stated, we add Poisson noise that would be typical of data collection scans in the 2- to 4-minute range, typically giving 0.1 to 0.5% (1σ) repeatability and about 15 to 20 wafer sites per hour, using a microfocus x-ray tube. As explained above, and in more detail in Chapter 5, there is a trade-off between throughput and repeatability. Typically, 75 wafer sites can be processed by XRR in an hour if repeatability is relaxed to 0.33%, and as many as 150 if it is relaxed to 0.65%.

The limits are continually being extended by the development of high-brightness sources and the optics to utilize them efficiently. The task of the tool design engineer is to achieve the best compromise, and this must be done with an understanding of the real needs of the fabrication line.

2.2 Dielectrics and Metals

In general we measure thin layers by interference effects, whether in diffraction or reflectivity. These work well up to approximately 0.5 μm thickness (~1 μm for very low density materials), after which absorption in the upper layer and the closeness of the fringes renders them impractical. Other x-ray methods, based on intensity measurement in diffraction or fluorescence, can also be useful, especially for thicker layers. These are not metrologically traceable and must depend on separate calibration.

2.2.1 Interferometric Methods

A common need is for the measurement of simple single-layer structures such as gate oxide films, Cu interconnects, and high-*k* and low-*k* dielectric films. These are measured by XRR. Since it is a grazing incidence technique, small spots are impractical, but it is feasible to measure a narrow stripe, for example, within a scribe line.

XRR measures thickness by means of fringe spacing. The fringes are not quite uniformly spaced, but are compressed slightly toward low angles (technically they are a weak chirp function). The spacing is calculated exactly from the equation for the fringe number *m*:

$$\theta_m = \sqrt{\left[\frac{m\lambda}{2d}\right]^2 + \theta_c^2} \tag{2.1a}$$

$$\theta_m = \sqrt{\left[\frac{\left(m + \frac{1}{2}\right)\lambda}{2d}\right]^2 + \theta^2} \tag{2.1b}$$

for the cases where the layer electron density is less or greater than that of the substrate, respectively, and in which θ_m is the incident angle and θ_c is the critical angle for total external reflection. Critical angle depends on electron density, which is directly related to physical density and chemical formula, so the density is also determined by the position of the critical angle and the detailed spacing of the fringes. This is an enormous advantage of XRR over optical methods, in which optical constants may be very different for the thin film and the bulk. In x-ray reflectivity, the refractive index corrections to the data are not only small but are contained in the data set (critical angle) even if the composition is unknown. This is a great help in process development.

There must be sufficient contrast (relative to the system noise) in the fringes, for which there are two conditions:

1. The surfaces and interfaces must be smooth enough to give good reflectivity.
2. There must be sufficient electron density difference between adjacent layers to give contrast.

The first condition, implying typically <3 nm root mean squared (rms) roughness, is usually satisfied in semiconductor thin films. As in optical interferometry, the best fringe contrast is achieved by approximately equal reflectivity at each surface/interface that is reflecting. Polysilicon on a single crystal silicon is invisible. Otherwise, the most common difficult case in common semiconductor materials is that of silicon oxide on silicon, because of the similar electron densities, but even these may usually be measured. Examples from semiconductor metrology are illustrated in Figure 2.1.

If the layer is too thick, then the transmitted wave is strongly absorbed and again contrast becomes insufficient. Also, at large thickness the fringes become very close together, and very high resolution optics are needed to

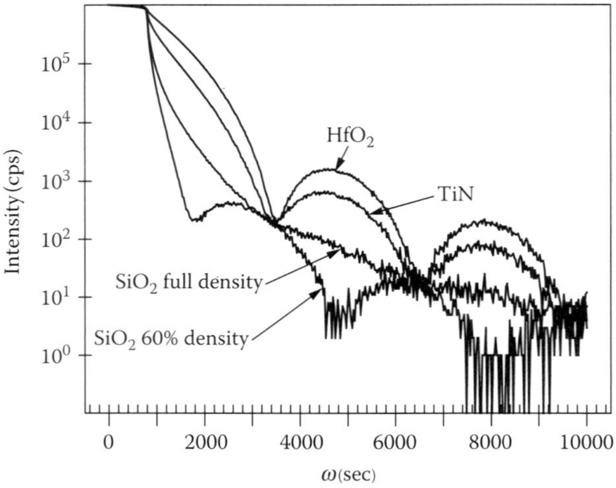

FIGURE 2.1
Effect of material. XRR curves from SiO_2 at 100% density, SiO_2 at 60% density, and HfO_2, each 5 nm thick.

measure the fringes. Examples are shown in Figure 2.2. The practical upper limit is 0.5 to 1 μm, depending on the film density.

The lower limit is likely to depend on the film structure, but 1 nm is normally practical. Measurement of thin layers depends on collecting data up to large scattering angles (2 to 4°), where the signal is weak and the detector noise important. Simulations show that if the layer is uniform, it is quite easy to measure 1-nm layers of, say, high-*k* dielectrics. However, experience tells that very thin layers are often nonuniform and may show significant grading near the surface. This is shown in Figure 2.3, which also shows

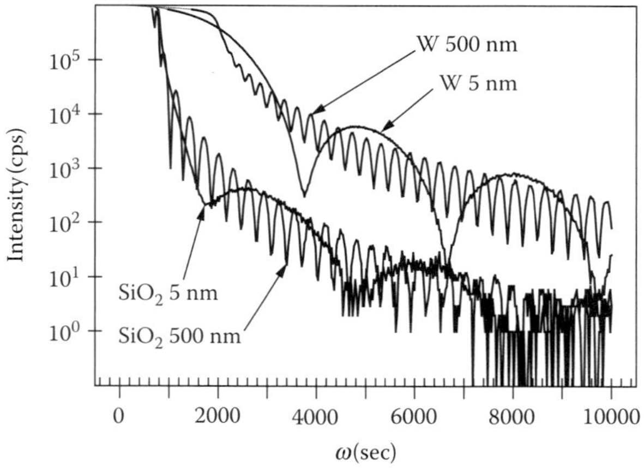

FIGURE 2.2
Effect of increasing thickness. XRR curves from 5 and 500 nm of SiO_2 at 60% density and of W.

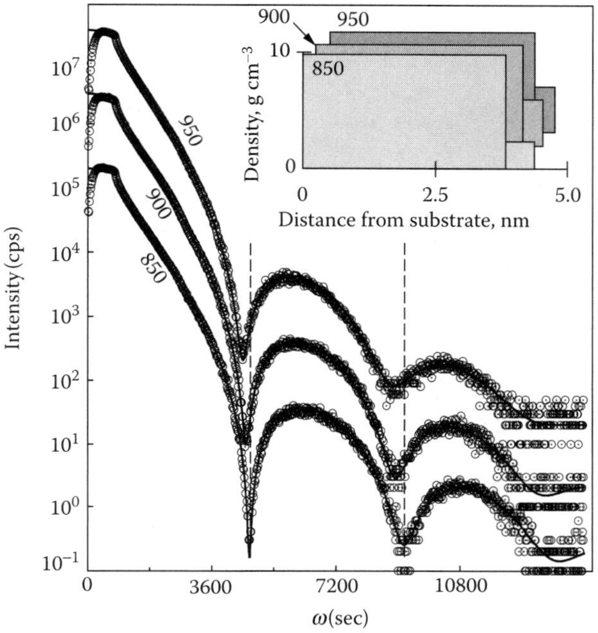

FIGURE 2.3
XRR on deposited and annealed HfO_2. Rapid thermal annealing was performed at 850, 900, and 950°C for 20 sec. The dotted lines are drawn at the position of the first minimum in the 850°C curve. The inset shows the density of the dielectric, modeled as two layers, as a function of distance from the substrate. (Courtesy R. Matyi, NIST.)

that the grade (modeled approximately as a single lower-density layer) changes after annealing treatments.

Thin buried layers may also be measured if the covering layer is not too thick. In fact, they may be simpler to measure than thin surface layers, since they are sensitive to both the phase and the contrast of the fringes arising from the covering layer. This may be measured at lower angles than the fringes from the thin layer itself (Figure 2.4). This is similar to heterodyne interferometry or amplification, in which a small signal is detected not directly but by its influence on a larger signal.

However, if the electron densities of the thin layer and the covering layer are very similar, the metrology is more difficult and depends on a good material model. For example, TaN/Ta barrier layers deposited by automatic layer description (ALD) and by sputtering have different structures (ALD has a much sharper interface), and these must be taken account of in the modeling for the metrology to be accurate.

2.2.2 Intensity Methods

While the traceable interference methods form the basis of x-ray metrology, the same tool may usually be used to make intensity measurements in either

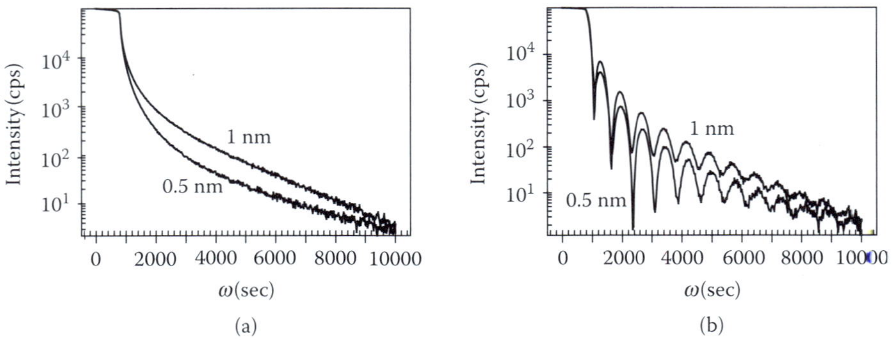

FIGURE 2.4
Measuring a thin buried layer. (a) HfO_2 layers 0.5 and 1 nm thick on a Si substrate. The thicker layer reflects more strongly, but the interference fringe period is too long to be easily visible. (b) As (a), but with a 20-nm Al layer on top of the dielectric. The fringe modulation will give greater accuracy to the measurement.

diffraction or x-ray fluorescence. There are times when this is useful, for both polycrystalline and amorphous layers.

As an example, damascene copper layers are usually too thick to be measured by XRR interference. They are strongly textured crystallographically, consisting of small grains that cluster about the [111] direction orientation normal to the surface. Measurement of the intensity of the 111 reflection, integrated over a small range of angles about axes parallel to the surface, gives a fast and repeatable measure of the layer thickness. This method requires calibration against a film of known thickness. There will be nonlinearity as absorption effects begin to be significant. The method described is valid up to about 10 μm, ample for the range of films encountered in semiconductor processing. If the material has a weak texture, a large angular range will be required for the integration.

XRF intensities are not affected by the crystalline state or orientation of the material. The XRF signal is linear with the quantity of an element irradiated by the beam, after making corrections for absorption of the incident and exit beams and for secondary fluorescence. If the density is constant and known, this may be calibrated with a standard film (as discussed in Chapter 1) and converted into a repeatable thickness measurement. Again, nonlinearity is expected. However, density variations are common in thin layers, especially near the surface, and XRF is better reserved for the measurement of layers greater than ~1 μm thick. It is necessary to correct the results for secondary fluorescence when the fluorescence from one analyte can be excited by the fluorescence from another. For example, consider that copper and iron are both present in an alloy that is being irradiated by MoK radiation. The copper will fluoresce only from the Mo radiation, but the iron, being of lower atomic number, will fluoresce both in the direct Mo radiation and in the fluorescent radiation from the copper. The iterative procedure for performing analysis is shown in the inset.

Fundamental parameter XRF analysis

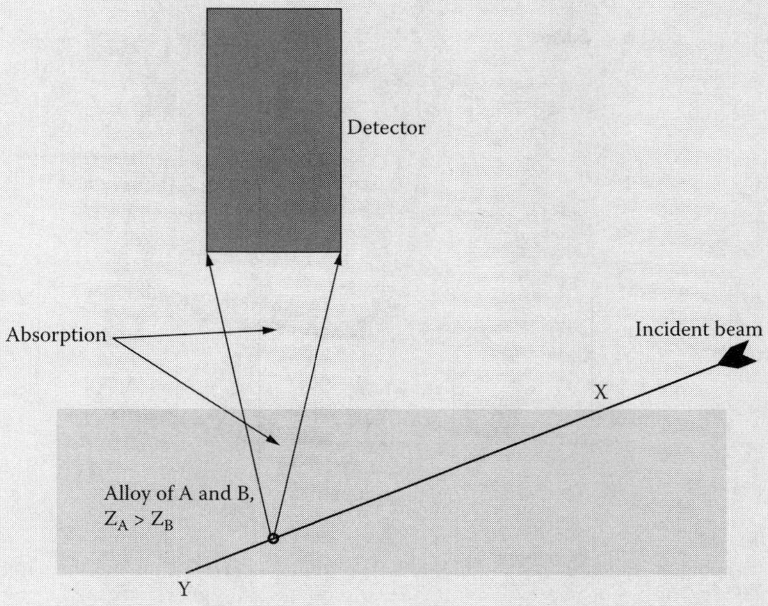

The emitted XRF intensity from an analyte can be calculated by multiplying the following factors together:

1. Absorption of beam entering the sample, down to depth of analyte
2. Number of analyte atoms in the beam
3. Efficiency of fluorescence (XRF cross-section)
4. Absorption of fluorescent beam leaving the sample

These may be calculated from the geometry of the measurement plus physical property databases. All the absorption factors and cross-sections are known for all the elements.

In addition there are factors that are common to a given experimental setup and exciting and fluorescing wavelengths. These may be calibrated by a single measurement on a 100% pure standard of the analyte:

1. Incident beam intensity
2. Detector aperture and efficiency
3. Absorption of emitted radiation in air and the detector window

Finally, there are cross-effects, dependent on the composition of the sample. If the sample is an alloy of A and B, with the atomic number Z of A sufficiently greater than that of B, fluorescent radiation from A atoms is generated along the path XY, and will excite further fluorescence from B atoms in the analyzed volume.

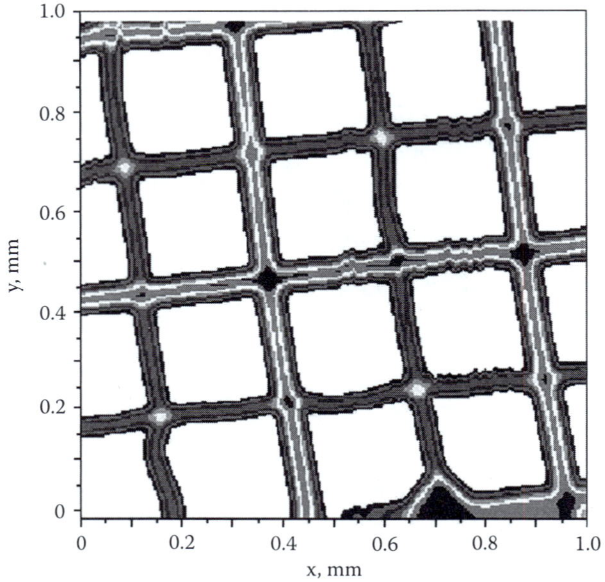

FIGURE 2.5
XRF mapping of grid bars from a TEM grid. The wider bars are 30 μm and the narrower bars 20 μm wide. The bright area arises from the CuKα fluorescent signal stimulated by focused Mo radiation.

Figure 2.5 shows an example of how, with a small, focused x-ray beam, the distribution of a particular element can be mapped across the wafer surface. The bright area arises from the CuKα fluorescent signal from a Cu TEM grid. With suitable x-ray optics, spatial resolution of better than 30 μm is achievable.

XRD can also be used to measure the thickness of layers typically a micrometer thick. The intensities of the Bragg peaks scales with thickness and by calibration against a known standard, the thickness can be deduced. It is important to note that it is the integrated intensity under the Bragg peak that should be measured, not the peak height. The peak width decreases as the thickness increases, and it can also be broadened by the presence of strain. Both effects are removed from the measurement if the area under a Bragg peak is taken as the metrological measure. When using this method, care should be taken to ensure that the texture of all the films being examined does not change; films with a different orientation distribution of crystallites give peaks of different relative heights. (Note that the texture of calibration sample must also be the same as the samples under examination.)

A further hazard in the use of integrated Bragg peak intensities for thickness metrology arises if the film is single crystal. While there is no problem for polycrystalline films such as sputtered interconnects, epitaxial films, such as $Al_xGa_{1-x}As$ of about 1-μm thickness, suffer what the old-time crystallographers referred to as extinction. Here, multiple scattering of the x-rays starts to become important, and depending critically on the perfection of the material, the film diffracts neither as an ideally imperfect nor an ideally perfect crystal. It may be impossible to distinguish whether changes in the integrated intensity under the Bragg peak arise from change in perfection or thickness. Some polycrystalline structures can also cause extinction effects, and XRD intensity measurements should therefore be used as a fingerprint method, not as a metrology.

2.3 Multiple Layers

When multiple layers are present, we obtain in principle a set of XRR fringes for every combination of two interfaces in the material. These add in amplitude, not intensity, so they cannot be deconvolved, though some information on individual layers can be gained from Fourier methods. The reflectivity profile rapidly becomes complicated, and intuitive fringe spacing ideas are inadequate for analysis. The merits of direct and modeling methods are discussed in detail in Chapter 13, but in our opinion it is unarguable that for complex structures the modeling approach is the only practicable one.

The layers must be uniform laterally within the size of the beam. Thus, although complete microcircuits may have very many layers, the lateral structure of the microcircuitry makes it impractical to measure the complete structure by XRR. The XRR is therefore performed on a characterization spot, which may be within a scribe line. The test structures that must be measured normally have only a few layers.

Metallic multilayers form the basis of a family of devices used for data storage and retrieval. The most important is the spin valve, which consists of a sequence of magnetic and nonmagnetic layers designed in such a way that the magnetization in one layer can be reversed easily. Reversal of the magnetization direction results in a large change in the resistance of the multilayer, the so-called giant magnetoresistance. Spin valves are now incorporated into read heads for computer hard discs and provide the means for reading high-density recorded data. Control of layer thickness to the angstrom level is necessary in the fabrication of spin valves by magnetron sputtering, and XRR is the key metrology in this area. Figure 2.6 shows an example of the specular reflectivity from a spin valve with structure Si/Sio_2 (substrate)/Ta/NiFe/CoFe/Cu/CoFe/NiMn/Ta (curve H). Although simulation of the interference fringes results in an accurate determination of the total transition metal layer thickness, the difference in x-ray scattering factor between Cu and the various Ni-Fe-Co-Mn alloys is very small, and the position of the interface between these two layers cannot be identified. To determine the thickness of each individual layer as the process was developed, it was necessary to perform XRR after growth of each layer as the stack was constructed. These are shown in curves A to G in Figure 2.6. The

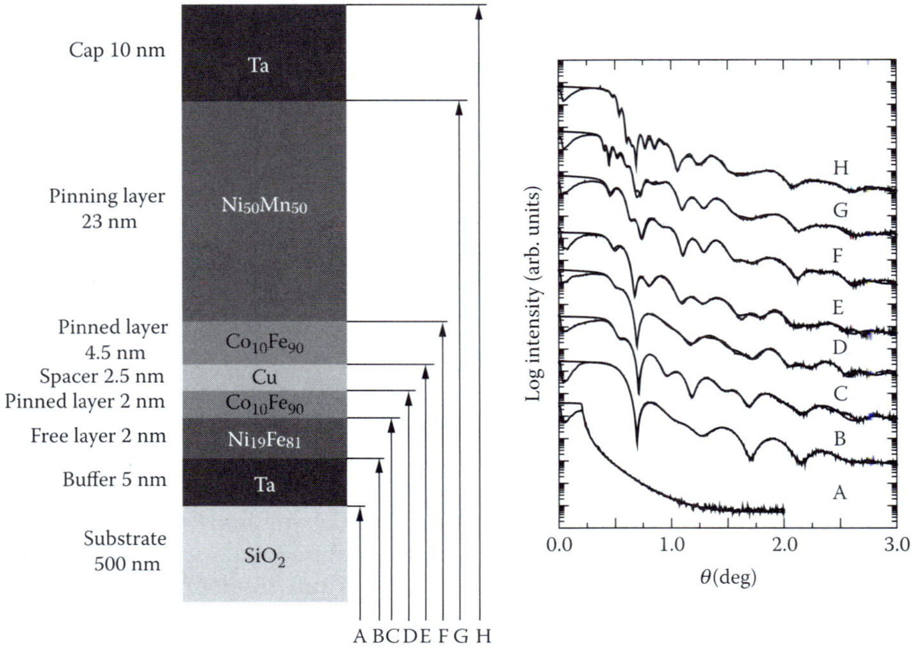

FIGURE 2.6
Determination of a spin valve structure by XRR after each stage of growth. The structure is shown on the left. Curves A to G show XRR after each stage of growth (identified on the schematic of the structure), and curve H is the XRR curve of the whole structure. Simulations (smooth lines) of the best fits obtained at each stage are superimposed on the experimental data. (After Brown, E. and Wormington, M., *Adv. X-Ray Anal.*, 44, 290, 2001. Grazing Incidence In-Plane Diffraction in the Laboratory, *Adv. X-Ray Anal.*, 47, © ICDD 2004. This material is used by permission of ICDD.)

simulation at each stage gave a very accurate determination of the overall structure.

A much greater magnetoresistance, and hence read sensitivity, can be obtained from devices in which there are still two magnetic layers, one in which the magnetization is pinned and one in which it is free to rotate in a small field, but now separated by an insulating layer. If the insulating layer is sufficiently thin, electrons can quantum mechanically tunnel between the magnetic layers. The tunneling probability is substantially different when the magnetization of the two magnetic layers is parallel or antiparallel, leading to a large tunneling magnetoresistance (TMR). The resistance change is much greater than for propagation through the nonmetallic (Cu) spacer layer in a spin valve, and therefore TMR devices are likely to form the basis for the next generation of read heads.

Figure 2.7 shows an example of the XRR from a typical TMR structure. The best-fit layer parameters are given in Table 2.1.

Multilayer structures of Co/Pd are potential storage media for high-density perpendicular magnetic recording. These are artificial superlattices in which the Co thickness must be typically 0.3 nm in order that the

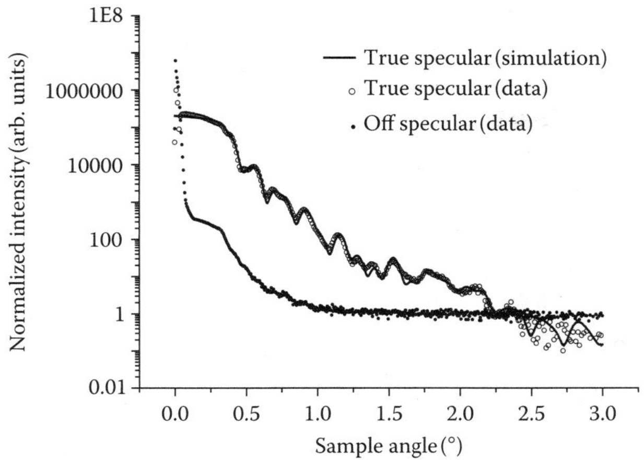

FIGURE 2.7
An example of the XRR from a typical TMR structure. The best-fit layer parameters are given in Table 2.1.

TABLE 2.1

The Best-Fit Layer Parameters for the Structure Whose XRR Data Are Shown in Figure 2.7

Layer	Thickness (nm)	Interface Width (nm)
Si	Substrate	0.65 ± 0.05
Al_2O_3	23.9 ± 0.5	0.34 ± 0.05
Co	12.4 ± 0.5	0.45 ± 0.05
Al_2O_3 (barrier)	3.12 ± 0.1	0.23 ± 0.05
NiFe	15.3 ± 0.5	0.21 ± 0.05
Oxide	0.31 ± 0.2	0.68 ± 0.05

magnetization orients normal to the film surface, the so-called condition for perpendicular anisotropy. A typical XRR profile for an eight-repeat sputtered Co/Pd multilayer is shown in Figure 2.8. The thickness of the Co + Pd layer is given by the separation of the superlattice Bragg peaks, and the relative peak heights are used to determine the individual peak thickness values. As the number of repeats increases in the artificial superlattice, the low-angle Bragg peaks sharpen. There is no abrupt change between a complex Keissig interference fringe pattern from a low number of repeats and the sharp Bragg peaks from many repeats. With increase in repeat number, some interference fringes strengthen at the expense of others and the Bragg peaks emerge at these positions. The advantage of working with many repeats is that simple interpretation can be used on the Bragg peaks to obtain approximate layer parameters without resorting to simulation and data fitting.

XRF is not applicable to superlattice structures (apart from obtaining the mean composition of the set of layers) but is useful for multiple (~6) layers

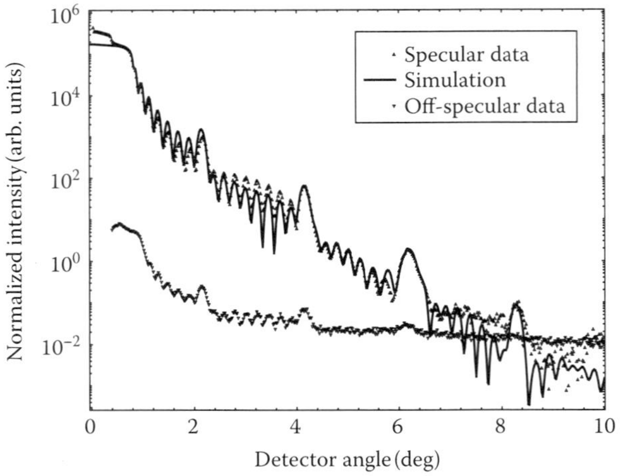

FIGURE 2.8
A typical XRR profile for an eight-repeat sputtered Co/Pd multilayer.

of relatively thick films. There are cross-effects, in that the intensity detected from an analyte is dependent not only on its own concentration but on that of all other analytes. The fundamental parameter method is used to calculate the absorption (in both incident and emitted beams) and secondary emission in each layer separately, and iteration is used to converge upon a solution. It is necessary to determine all the compositions and thicknesses of all the layers together.

2.4 Epitaxial Layers

2.4.1 Interferometric Methods

Since XRR is indifferent to the crystalline state, it may be used for epitaxial as well as for amorphous layers, with exactly the same methodology as for metals or dielectrics. An example is shown in Figure 2.9.

For good-quality crystalline layers, HRXRD may also be used for thickness metrology using interference fringes, as introduced in Chapter 1. This works up to about 1 μm. As with XRR, the thicker layers require higher-resolution optics, since at 1-μm thickness the fringe spacing is only about 20 arc sec. The only material-dependent factor in the calculation of thickness is the Bragg angle, which is measured during the process to an accuracy that does not affect the accuracy or repeatability of the final result.

Multiple epitaxial layers again produce fringes that add in amplitude and are best resolved using a modeling approach. Compound semiconductors provide the best examples of complex structures in which each layer or superlattice can be measured, as shown in Figure 2.10.

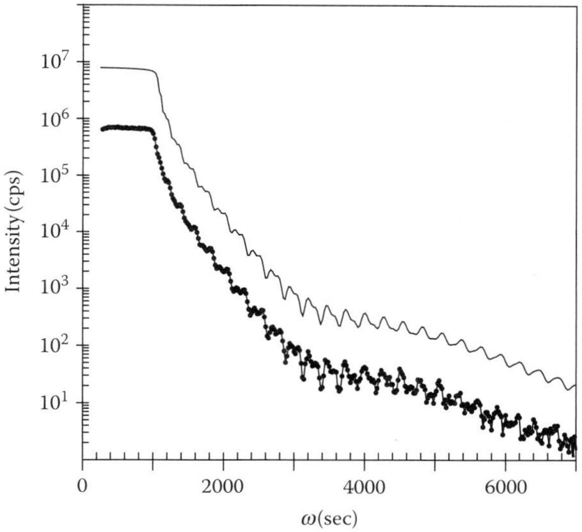

FIGURE 2.9
Specular reflectivity curve of a AlGaAs + GaAs bilayer on a GaAs substrate, together with a simulation from the best-fit model (displaced over a decade for clarity). The layers, which are closely matched and about 50 nm thick, cannot be properly resolved by HRXRD. The measurement by XRR is straightforward.

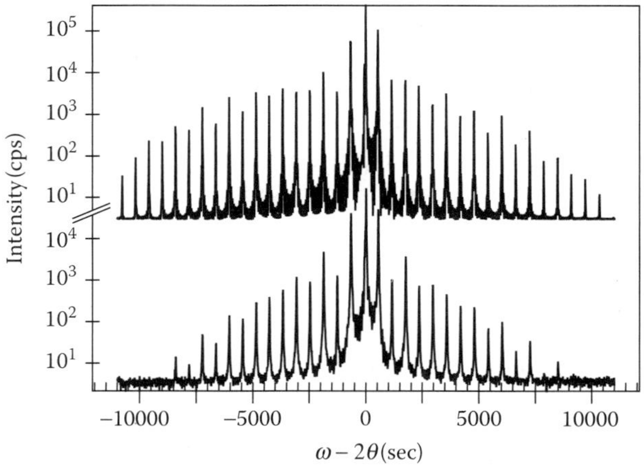

FIGURE 2.10
HRXRD curve from compound semiconductor superlattice. InAs substrate, with 30 repeats of (12.8 nm InAs/2.5 nm $InAs_{0.7}Sb_{0.3}$/12.8 nm InAs/2.5 nm $Al_{0.25}In_{0.75}As$). (Courtesy B. Dutta, IMEC.)

The layers need not be uniform composition vertically, and in fact, graded layers are important in certain heterojunction bipolar transistor (HBT) or pseudomorphic-high-electron-mobility-transistor (pHEMT) structures. This

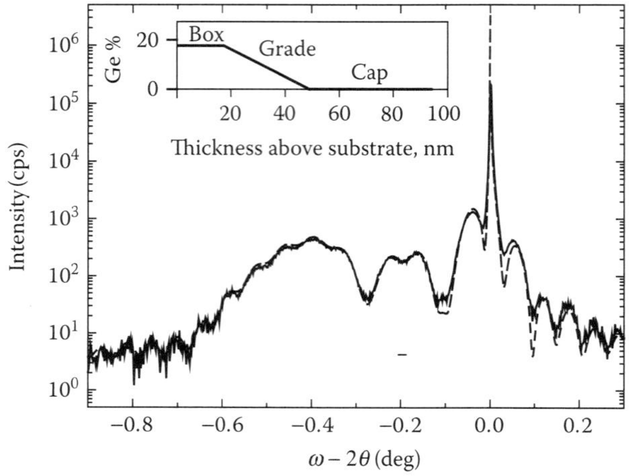

FIGURE 2.11
HRXRD data from a SiGe graded composition structure (solid line) and the simulated curve (dashed line) from the best-fit model. The inset shows the structure that was deduced.

is handled in modeling by dividing the layer up into uniform lamellae, i.e., a staircase approximation to the grade. An example is shown in Figure 2.11.

2.4.2 Intensity Methods

Intensity methods may also be used for epitaxial layers such as relaxed SiGe layers, in which the defects destroy the fringe contrast. The integrated intensity of the layer peak is compared with that of the substrate peak, and a calibration curve is generated from a modeling program or from experimental data. It is important to use integrated, rather than peak, intensities, since any relaxation or curvature of the wafer will modify the peak intensities but have a second-order effect on the integrated intensity. However, the type and distribution of defects should be similar between the measured layer and the calibration or modeled layer. Otherwise, differences in the type of x-ray scattering (extinction effects) may introduce significant errors for layers greater than about 0.5 μm in thickness.

2.4.3 Small Measurement Spots

HRXRD is usually taken at an incidence angle of 30 to 40° or even higher. The spot on the specimen is elongated by a factor of less than 2. It is therefore possible to perform HRXRD on patterned wafers, using optical alignment and pattern recognition techniques to locate the x-ray spot on a particular characterization spot built into the die pattern, or even within the scribe line. Throughput is lower than on a blanket wafer, because of the small scattering volume, but the measurement is made on material grown in identical

	Site 1	Site 2	Site 3	Site 4
SiGe t (nm)	58.0	56.6	56.3	58.5
SiGe X (%)	22.97	22.69	22.71	23.02

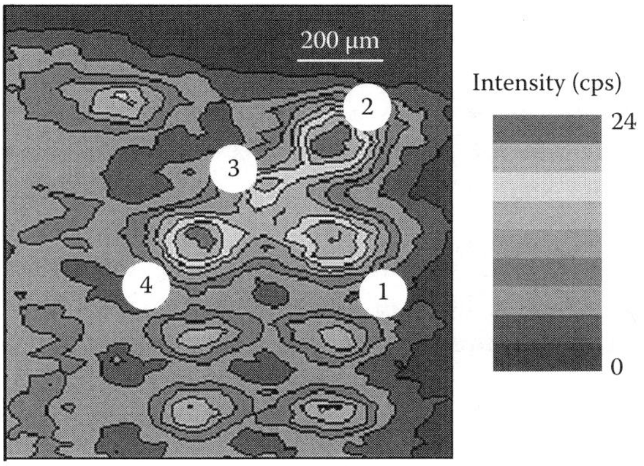

FIGURE 2.12
HRXRD data taken using a 100 × 100 μm beam. The specimen contains a number of test pads 200 μm in size. An HRXRD scan was taken of a pad, then the diffractometer was set on the SiGe peak and an area map taken to show the pads. HRXRD scans were then taken at the numbered test pads and the structure determined. The table shows the small variations in thickness and Ge content that were identified. Relaxation scans showed that the SiGe was fully strained in the test pads.

conditions (selected area epitaxy) to devices. An example of mapping with a finely focused x-ray beam is shown in Figure 2.12.

2.4.4 Comparison of XRR and XRD for Epitaxial Thickness Metrology

We may now address the question of when one uses XRR and when one uses HRXRD for measurements of thickness in epitaxial layers. The principles of the difference are:

- XRR is insensitive to strain; it is only sensitive to electron density.
- XRR is a grazing incidence method.
- HRXRD is very sensitive to strain differences between layers.
- HRXRD is a large-angle incidence method.

XRR should be considered when:

- Blanket wafers or scribe lines can be measured.
- Strains are large and inhomogeneous, e.g., in thick, heavily relaxed layers.

- Strains are very similar between layers, e.g., lattice-matched quaternary layers in compound semiconductors.

HRXRD is the better choice when:

- Spots on patterned wafers must be measured.
- Electron density differences are small, e.g., strained silicon.
- Strain differences outweigh electron density differences (e.g., most SiGe thin-layer structures).

It is advisable and quick to run simulations of model structures in cases that are not obvious. This is usually necessary in any case, to calculate the throughput and repeatability that can be expected for a given metrology.

2.5 Summary

- XRR will measure thicknesses from approximately 1 nm to 0.5 to 1 μm in most materials in which interface roughnesses are less than about 3 nm.
- XRR cannot be used on a small spot measurement but can be used in a scribe line between dies.
- XRR can be used for any type of thin-film material, provided the electron density is not very close to that of the neighboring layers.
- HRXRD can be used for thicknesses of uniform or graded epitaxial layers, from approximately 2 nm to several micrometers.
- Both XRR and HRXRD can be used for multiple-layer structures.
- Both XRR and HRXRD have current throughputs of up to 10 to 20 data points per hour at 0.1% 1σ, rising to 150 per hour at 0.65% 1σ.
- XRF is best used for thick films of single or multiple layers, and as supplementary information when films have very similar electron densities.

3

Composition and Phase Metrology

3.1 Introduction

There are many cases in which the precise composition of a thin film radically affects its properties. Sometimes, the composition is stable since the material is an element or stable stoichiometric compound, but even then there can be doubts. Has that stoichiometry actually been achieved in manufacture? Is that pure element contaminated? For alloys, such as SiGe, the whole electrical performance of the layer or of the next strained silicon epilayer will depend upon the composition. It is a parameter that must be measured and controlled and is a particular problem when films get very thin. The classical analytical methods such as optical spectroscopy or conventional x-ray fluorescence (XRF) do not then work well. Optical spectra may be very different in very thin films, and though XRF spectra will be true to the volume sampled, the probing beam will often excite fluorescence in several layers.

X-ray methods give three means of measurement of composition:

1. Lattice parameter methods: The change of lattice parameter with composition is calibrated or assumed to be linear. This method is applicable to single or polycrystals and is highly reproducible, but for the best accuracy calibration is required.
2. Fluorescence methods: The intensity of a fluorescent signal unique to an element is measured. This method is (almost) unambiguous in identification of a given element. Calibration is required for accuracy, and grazing incidence methods are required to sort out the distribution of elements between different layers.
3. Scattering power methods: The strength of either reflectivity or diffraction is proportional to the number of electrons in the atomic type that is doing the scattering. This can be calculated from first principles, but results in a mean electron density for the material. Further information is usually required for analysis.

It is often sensible to combine the XRF methods with techniques such as diffraction or reflectivity in order to get unambiguous information.

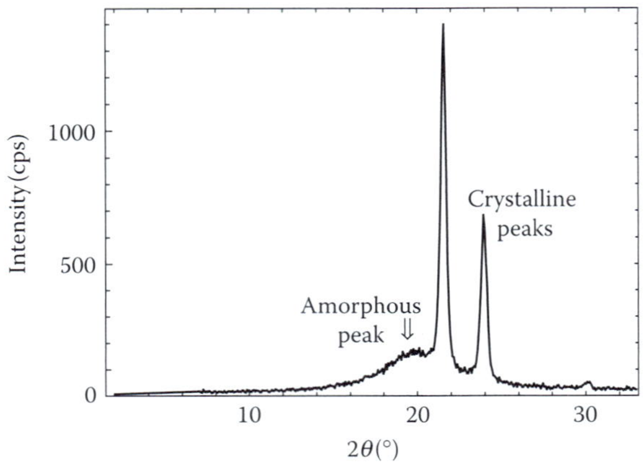

FIGURE 3.1
X-ray diffraction pattern from a high-density polyethylene sample with just over 50% crystalline fraction.

3.2 Amorphous Films

Although x-ray diffraction (XRD) gives almost no information on the composition of amorphous films, it does provide a hugely important method for determining whether crystallization has occurred. In bulk polymers, for example, measurement of the intensity under the sharp Bragg peaks from the crystalline fraction, compared with that under the broad scattering hump from the amorphous material, gives directly the crystalline fraction. Figure 3.1 shows an example of the scattered intensity from a piece of high-density polyethylene with just over 50% crystalline fraction. This has been applied, for example, to the detection of crystallization in high-*k* dielectrics only 2 nm thick. It is also an issue in metal gate layers some 10 nm thick. Crystallization in these materials is deleterious to their required electrical and sometimes physical properties.

X-ray reflectivity (XRR) is sensitive to the electron density only, and this is a combination of physical density and chemical composition. An important example of the application of XRR is in the determination of the sp^3 fraction (i.e., the fraction of diamond-like bonding) in amorphous diamond-like carbon (DLC). The tribological properties of DLCs make them of great importance in covering magnetic recording media. Plasma deposition techniques are difficult to control, and metrology is necessary in the preparation of DLC films. The sp^3 fraction is linked directly to the physical density, and measurement of the critical angle from the XRR provides a unique method (Figure 3.2).

XRR is of no use for analyzing the composition of an unknown material, but can be used, for example, to find the boundary between metal and nitride in a diffusion barrier layer.

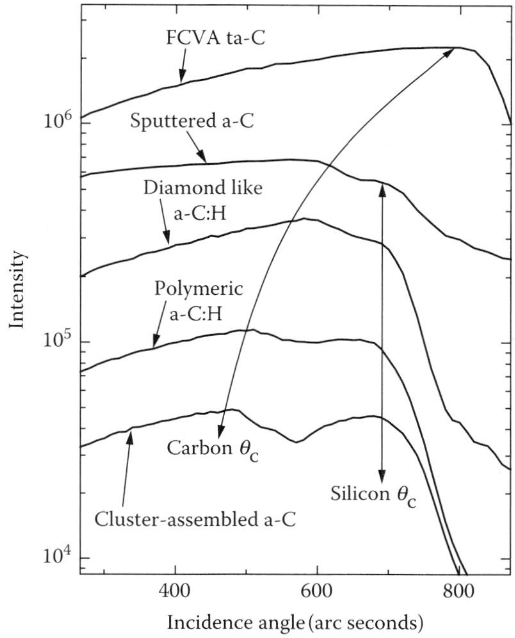

FIGURE 3.2
Experimental x-ray reflectivity spectra for diamond-like amorphous-C:H, polymeric a-C:H, and cluster-assembled a-C films. Their electron density is lower than that of the silicon substrate, so a double critical angle appears, as indicated. The top line is the reflectivity curve of a tetrahedral amorphous-C FCVA (Filtered Cathodic Vacuum Arc) film, whose density is greater than in Si; in this case, a single critical angle is detected. (From Tanner, B.K. et al., *Mater. Res. Soc. Symp. Proc.*, 615, G1.2.1–G.1.2.12, 2000. With permission.)

Where we have an amorphous ternary layer, however, neither XRD nor XRR will help. An example is the compositional metrology of metal gate layers. These are amorphous ternary compounds, designed to lower the work function of the gate but still be compatible with manufacturing processes such as etching. The correct composition is crucial to these functions. XRF will measure the ratios of the constituent elements with ease, with calibration only from films or bulk samples of the pure constituent elements. Since the gate metals are analyzed straight after deposition, they are normally measured as a single layer. They are thick enough (typically ~10 nm) to be measured in normal incidence x-rays.

The penetration of x-ray waves during grazing incidence methods (Figure 1.1) is low, especially below the critical angle, and is controlled by incident angle. If the sample contains multiple layers and they happen to be arranged in order of decreasing density from the bottom, then the critical angles are reached in succession, and there is very good concentration of the beam to the layers in turn. This may be exploited to tune the sensitivity of x-ray fluorescence to individual layers. Again, it is best to model the complete fluorescence as a function of incident angle. This can be done with a

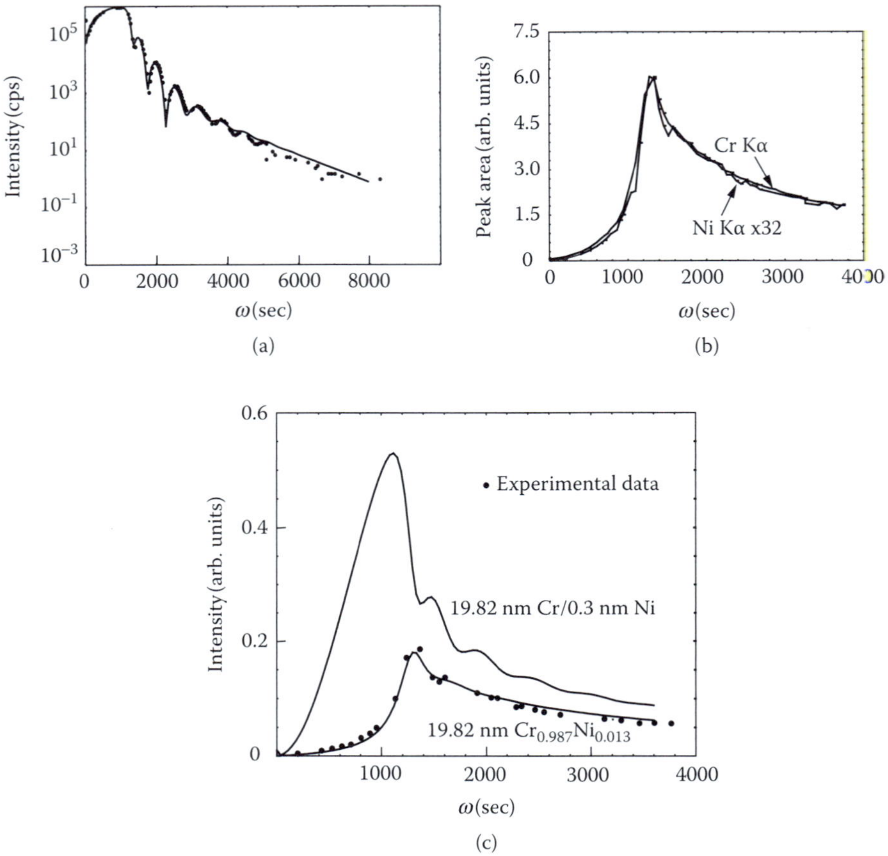

FIGURE 3.3
(a) Specular reflectivity (XRR) from a nominally 20-nm-thick Cr film on Si. The best fit to the data gives 19.8 ± 0.1 nm for the thickness. (b) Variation of the integrated intensity under the CrKα and NiKα fluorescence peaks as a function of incidence beam angle. (c) Fit of NiKα data to models of Ni at the surface and uniformly distributed through the film. (From Tanner, B.K. et al., *Mater. Res. Soc. Symp. Proc.*, 615, G1.2.1–G.1.2.12, 2000.)

methodology similar to that for the calculation of XRR. Figure 3.3a shows the XRR profile of a nominally 20-nm-thick chromium layer grown by ultra-high vacuum (UHV) evaporation on a silicon substrate. However, observation of the XRF spectrum reveals that there is about 1% contamination of Ni in the film. Figure 3.3b shows the variation of the integrated intensity under the CrKα and NiKα fluorescence lines as a function of incidence angle. The shapes of the two curves are identical, indicating that the Ni is distributed uniformly through the Cr layer. Figure 3.3c shows the Cr angular data fitted to the curves simulated for two models: (1) where the Ni contaminant is entirely on the surface and (2) where the Ni is evenly distributed in an alloy with the Cr. It is evident that there is an excellent fit to a model in which it is assumed that the whole Cr layer is contaminated with 1.3% Ni. If the

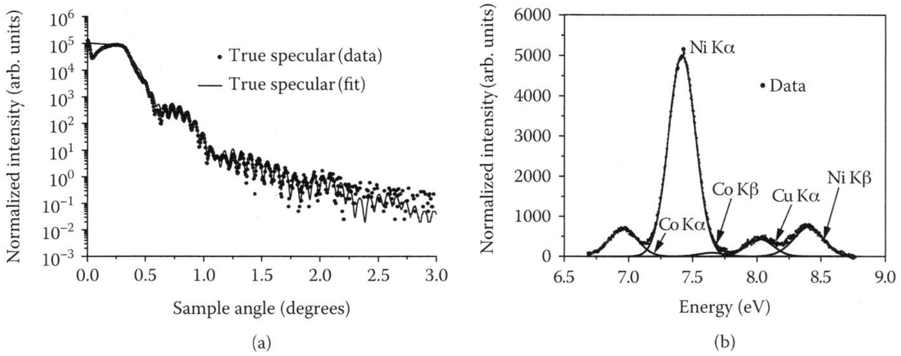

FIGURE 3.4
(a) Specular reflectivity from a spin valve of nominal composition SiO_2/NiO(50 nm)/Co (2.5 nm)/Cu (2.1 nm)/Co (3 nm). Table 3.1 shows the best fit to the data. No feature arising from the Co/Cu interface can be detected. (b) GIXRF from a magnetic multilayer showing the Co, Ni, and Cu fluorescence peaks. From the relative integrated intensity under the Co and Cu peaks, the ratio of the Cu and total Co layer thickness can be determined. (Courtesy Dr. J.D.R. Buchanan, Ph.D. thesis, Durham University, 2003.)

TABLE 3.1
Material Parameters Deduced from the XRR Data of Figure 3.4

Layer	Thickness (nm)	Interface Width (nm)
Co oxide	0.86 ± 0.5	0.75 ± 0.1
Co-Cu-Co	6.93 ± 0.5	0.48 ± 0.1
NiO	49.9 ± 0.5	0.17 ± 0.1
SiO_2	—	0.45 ± 0.1

layers are not arranged in order of density, then the penetration is still controlled by the angle of incidence, but some sensitivity is lost.

The grazing incidence x-ray fluorescence (GIXRF) method is particularly useful for distinguishing layers that have very similar electron densities, and hence reflectivities, but different compositions. This is not so common in silicon-based semiconductors, but often happens in magnetic multilayers. An example of XRR and GIXRF on a magnetic multilayer known as a spin valve, of nominal structure Si/SiO_2/NiO (50 nm)/Co (2.5 nm)/Cu (2.1 nm)/Co (3 nm), is shown in Figure 3.4. The XRR profile (a) shows high-frequency fringes associated with the 500-Å NiO pinning layer and a longer oscillation period arising from the Co-Cu-Co layers that form the magnetic switching layers. No feature is visible in the curve associated with the Co/Cu interface, and the thickness of the Cu spacer layer cannot be determined. Use of GIXRF in combination with XRR enables the thickness of the Cu to be determined. The GIXRF spectrum shows clearly resolved peaks from the Co, Ni, and Cu, and from the relative integrated intensities under the peaks, the relative thickness of the Cu and Co layers can be determined. As the total thickness of the Cu and Co layers is known from the XRR, the Cu thickness can be determined uniquely to a precision below 0.1 nm.

GIXRF can also be used in lattice-matched quaternary semiconductors, which have similar diffraction properties between layers, to distinguish the layers and hence obtain both composition and thickness. The GIXRF on its own would not be sufficient, since it cannot distinguish between thick dilute layers or thin concentrated layers, for a given analyte. This is an example of co-minimization of two separate effects (XRR and XRF), which gives much greater certainty in the measurement than either alone.

3.3 Polycrystalline Films

X-ray methods are much more powerful for crystalline materials. All polycrystalline materials give a unique diffraction pattern (with the usual caveats about sufficient signal/noise and resolution). Indeed, the peak positions alone are normally sufficient to identify the crystal, using the International Center for Diffraction Data (ICDD) database. If the grains are randomly oriented, or oriented with a known texture, then the intensities can also be predicted accurately. Furthermore, polycrystal diffraction patterns from mixtures of phases are simply additive in intensity. Patterns can be deconvolved to extract overlapping peaks, and quantitative analysis can be performed by intensity measurements. Reasonable patterns from polycrystalline thin films can usually be obtained down to thicknesses of 1 to 2 nm.

This may represent an ideal situation, since thin polycrystalline films often have some unknown texture that distorts the intensities. However, peak positions are not affected by texture, and in semiconductor metrology the issue is often to monitor deviations from a very stable and consistent process. An example is the composition of a metal nitride produced in a chemical vapor deposition (CVD) reactor. XRD will easily distinguish between different nitrides as long as they are crystalline. This is important for the efficacy of the diffusion barrier layers that are laid down before the copper layer in back-end processing.

Another application is to alloy content. While single-crystal SiGe is best analyzed by high-resolution x-ray diffraction (HRXRD) methods, analysis of polycrystalline SiGe is sometimes important. Silicon and germanium form a complete series of solid solutions, with about 2% difference in lattice parameter from one extreme to the other, so the composition analysis follows from the peak positions. Peaks can be located to about 0.01°, so the method gives approximately 0.5% precision.

Figure 3.5 shows the XRD patterns from two films of CdS on glass substrates, one predominantly in the cubic and one in the hexagonal phase. The presence or absence of certain specific peaks provides a qualitative indication

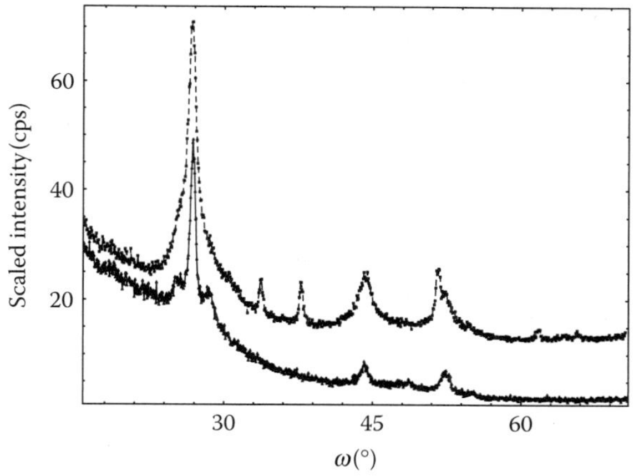

FIGURE 3.5
XRD patterns from two films of CdS on glass substrates. The pattern of peaks is a fingerprint of the structure of the phase of the CdS.

of the phase; the relative peak heights provide quantitative information on the relative fractions of the two phases.

3.4 Wafers and Epitaxial Films

Composition is determined in these materials by measuring the Bragg angle, and hence the lattice parameter. Thus, Bragg angle is the direct measurement, which may be either relative to a substrate or absolute. To derive the composition from the lattice parameter, we need one further piece of information for a bulk material, namely, the variation of lattice parameter with composition. For an epilayer, we also need to know the distortions between the epilayer in its coherent epitaxial form and in its free or relaxed state.

3.4.1 Variation of Lattice Parameter with Composition: Vegard's Law

Vegard's law simply states that the lattice parameter of a solid solution alloy will be given by a linear dependence of lattice parameter on composition, following a line drawn between the values for the pure constituents. Vegard's law was originally proposed for ionic salt pairs, e.g., KCl-KBr, but has been widely investigated for metals, in which it does not work too well, and widely assumed for III-V semiconductors. It is based on elastic interactions between atoms, and is thus reasonable when electronic interactions in the alloy series are very similar. Calibration curves are available for some alloys, most notably Si-Ge, and deviations from linearity are usually about 2% maximum.

TABLE 3.2

Dopant Calibration Factors

Dopant	To Convert True Strain to Dopant Concentration: $N = -\frac{\varepsilon_0}{\beta}$	To Convert True Strain to at%: $at\% = \frac{N}{N_{Si}} \times 100$
B	$N_B = -\frac{\varepsilon_0}{5.19 \times 10^{-24}} cm^{-3}$	$at\% = -\varepsilon_0(ppm) \times 0.000386$
C	$N_C = -\frac{\varepsilon_0}{8.91 \times 10^{-24}} cm^{-3}$	$at\% = -\varepsilon_0(ppm) \times 0.000225$
Sb	$N_{Sb} = \frac{\varepsilon_0}{1.10 \times 10^{-23}} cm^{-3}$	$at\% = \varepsilon_0(ppm) \times 0.000182$

The case of doping needs special consideration. For materials such as carbon on silicon, it does not make sense to interpolate linearly between the lattice parameters of silicon and diamond, and even less sense to construct a diamond-like boron, nitrogen, or antimony, all common dopants. Though the variation of lattice parameter with dopant composition is sufficiently linear in the small concentration ranges usually found, the effect of a dopant on lattice parameter must always be calibrated. Such calibrations do exist for carbon, boron, and antimony in silicon and are shown in Table 3.2, from the best available literature sources.

3.4.2 Coherency Distortion in Epilayers

This distortion is often called tetragonal distortion. The term arises because the materials used for semiconductor epilayers are usually cubic, and constraining them to fit coherently onto a cubic substrate of slightly different lattice parameter results in an extension or contraction normal to the interface. This makes them tetragonal. More complex distortions may arise if the substrate surface is not close to a cube face of the crystal structure, but these may usually be ignored. The practical issue is that in order to calculate and allow for this distortion, knowledge of the Poisson ratio of the material is necessary. These are available for silicon, germanium, and almost all III-V and II-VI compound semiconductors, but values for alloys are only obtained by linear interpolation akin to Vegard's law. It is not known for certain what errors are likely in this interpolation, but composition analysis by this method agrees very well with those obtained by other methods for the few cases studied. For example, in one study the correlation found in a comparison with secondary ion mass spectometry (SIMS) measurements was over 99%, as seen in Figure 3.6.

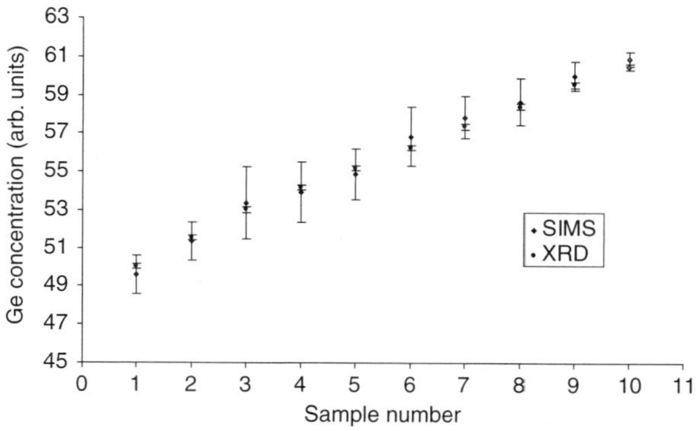

FIGURE 3.6
Comparison of SIMS and HRXRD results on a set of SiGe epitaxial wafers. The error bars for HRXRD are approximately one third of those of SIMS.

If the layer is not perfectly coherent but fully or partially relaxed, an additional measurement of lattice parameter parallel to the interface is required. Methods of performing this are discussed in Chapter 4, but in general it requires measurement of diffraction from an asymmetric plane, i.e., one not parallel to the interface.

Computational methods have been developed that determine both composition and relaxation, using the minimum number of measurements together with databases of lattice parameter, Poisson ratio, and any alloy calibration equations that are known. The overall accuracy of determination of composition of an epilayer, when calibration factors are known, is of the order 0.1 to 0.5% of value, with reproducibility typically some five times better than this.

3.4.3 Absolute Lattice Parameter Measurements

Absolute lattice parameter methods are usually applied to bulk wafers or boules and depend upon accurate measurement of the Bragg angle. The issue here is not only to measure an angle accurately, but to eliminate or compensate for the systematic errors that can arise due to zero error, misalignment, refractive index, wavelength uncertainty, etc. The most common error is tilt of the specimen relative to the diffracting planes of the beam conditioner; this broadens and displaces the peaks. The inset boxes show the methods used. We note that the triple-axis comparative method is very well based and easy to operate in a fab tool, since the lattice parameter of pure silicon is known to very high absolute accuracy (a few parts in 10^9). Even Czochralski silicon is known to 1 in 10^6 after compensation for oxygen and carbon content. In effect, the silicon is used to remove a large number of uncertainties at once. An accuracy of a few parts in 10^6 in a fab tool is realistic; any better requires special precautions, such as high-quality temperature control.

Measurement of lattice parameter
1. Single axis method (Bond)

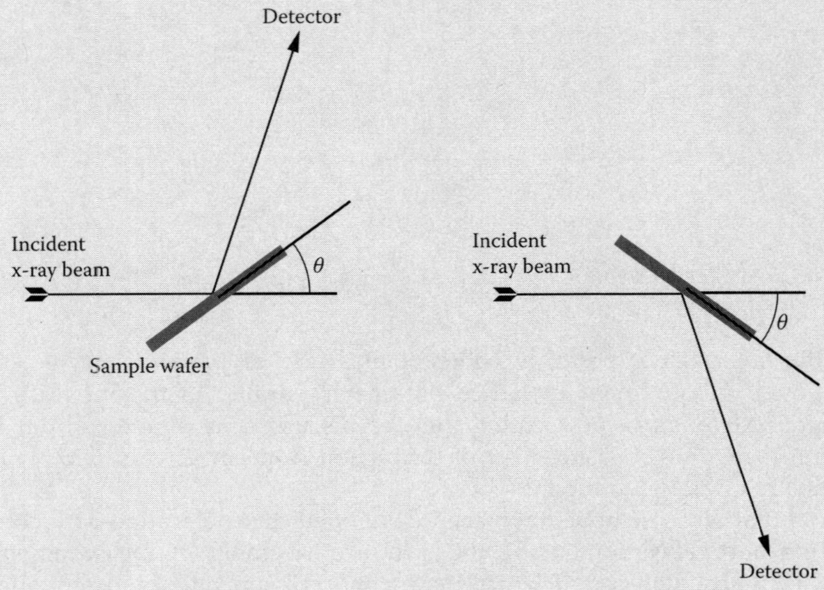

The Bragg angle is measured directly from the 2θ angle (the swing of the specimen between diffraction on both sides of zero) using open detector(s). Measurement on both sides of zero eliminates zero error.

Experimental reproducibility can be better than 0.01%.

Special comparators have been built[1] with accuracy better than 1 in 10^7, but have not been required thus far in a fab.

Compound semiconductor wafers such as $Cd_{1-x}Zn_xTe$, used as substrates for $Cd_{1-x}Hg_xTe$ infrared detectors, need to be grown with a very well defined Zn composition, normally about 4%. Absolute determination of the lattice parameter provides a straightforward metrological technique for control of the material composition to a precision of typically 0.1%. The agreement between x-ray metrology and optical techniques for band gap determination, such as photoreflectance, is extremely good[2] and within the measurement tolerances. With scanning stages, using an analyzer in the triple-axis mode, the lattice parameter may be mapped quite rapidly over the surface of the wafer. This is useful in compound semiconductors, since it reveals nonuniformities in the growth process. In Chapter 11 (Figure 11.11), we show an example of the variation in the composition of Ga in an $In_{1-x}Ga_xAs$ layer.

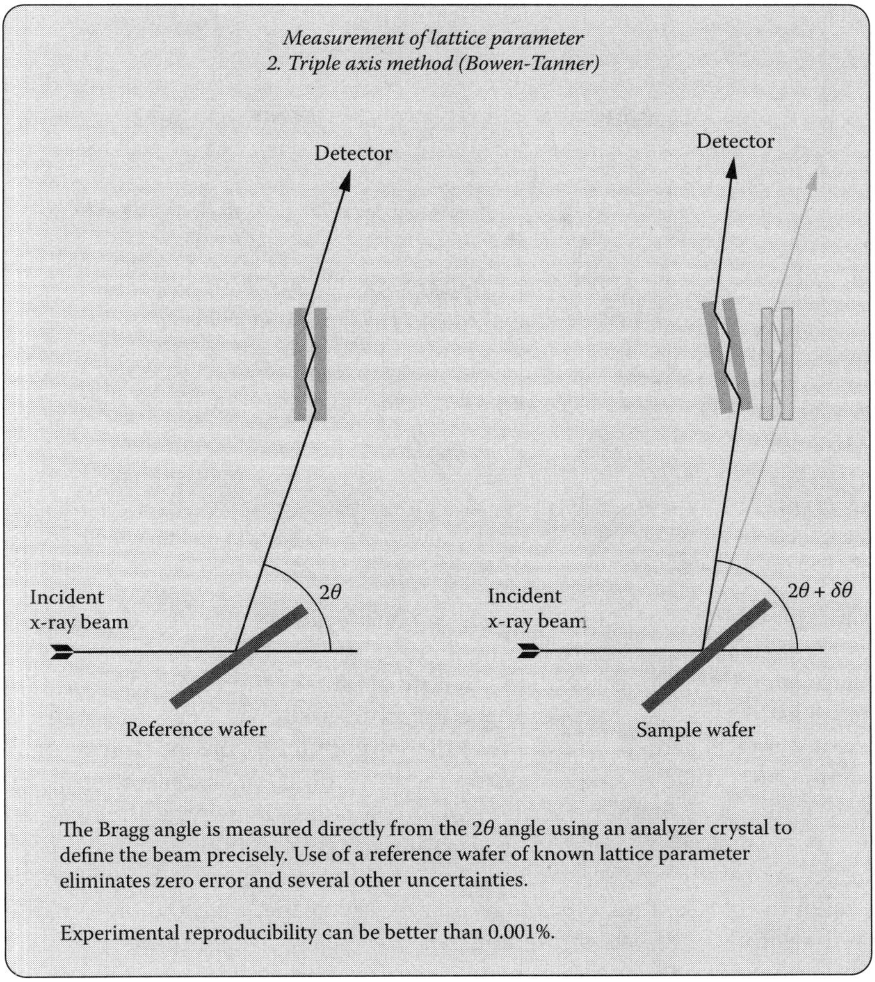

3.4.4 Relative Lattice Parameter Measurements

When an epilayer on a substrate is to be measured, we already have a good reference built in, and we may simply measure the splitting of the layer peak from the substrate. Examples are SiGe on Si, AlGaAs on GaAs. If the layer peak is not too thin, say >100 nm, a rocking curve (omega scan with open detector) is sufficient. However, this configuration is noisy, since it also collects diffuse scatter from defects in the material, as well as any spurious diffuse scatter from the tool. Signal/noise can be improved up to 50 times by use of a narrow slit (<0.1°), and this is necessary for materials such as heterojunction bipolar transistor (HBTs) and high electron mobility

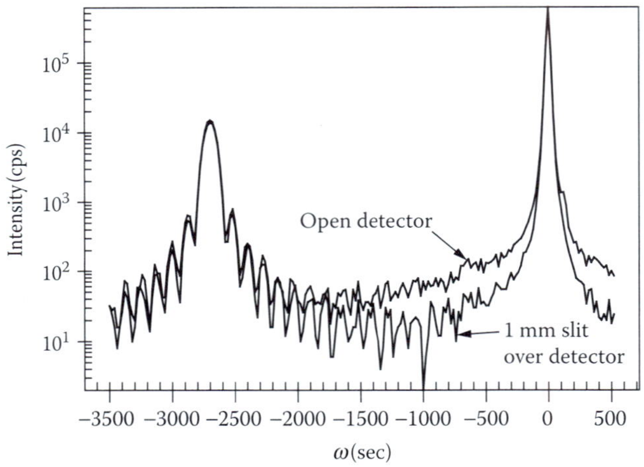

FIGURE 3.7
004 HRXRD ω-2θ scans of a GeSi layer on Si substrate, showing the effect of detector slits. Cu-sealed tube, 40 kV, 50 mA; step, 20 arc sec; count time, 0.5 sec per point. (Sample courtesy Hitachi-Kokusai, Japan.)

transistors (HEMTs), in which layers can be as thin as 5 nm. An omega-2theta scan is then used; see, for example, Figure 3.7.

However, there are precautions that must be taken in order to ensure that the measurement is accurate. If the epilayer is known to be coherent (not relaxed) and is grown on a surface that is parallel to the symmetry plane, usually (001), then a single measurement is sufficient. If relaxation is suspected, then additional measurements must be made using diffracting vectors (normals to diffracting planes) with a (large) component parallel to the surface. This will be discussed in the next chapter. If the substrate is not cut parallel to the (001) (or other high symmetry) plane, then the layer will be tilted relative to the substrate and it is necessary to determine this tilt to eliminate it from the measurement. There are two strategies:

1. Take two scans, with a rotation of 180° in ϕ (axis normal to the specimen surface) in between scans. Peaks due to the tilted layer will be displaced by equal and opposite amounts in the two settings, and so the tilts can be eliminated. This method doubles the data collection time.
2. Determine the axis of tilt. This requires three readings at different settings of ϕ, but will be the same for the whole wafer; or it may be known from the wafer manufacturing process. Then set the specimen to measure across the tilt, i.e., with the tilt axis coplanar with the incident and diffracted beam. The tilt then has no effect on the peak splitting. This method is faster if the orientation is known or if more than one point must be measured on the wafer. It is also the only method that allows direct comparison with modeled curves

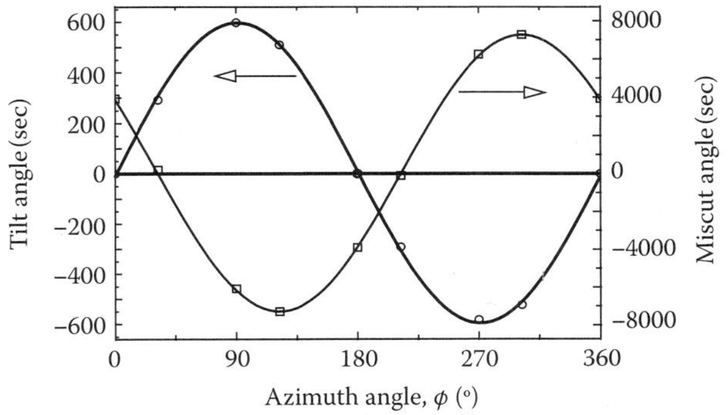

FIGURE 3.8
Determination of tilt in a layer grown on offcut substrate. The curves show the position of the peak from the substrate and layer, as a function of azimuth angle, and their fits to sine curves. The maximum amplitude of the curves is the tilt. The zero crossing points show where the tilt axis is perpendicular to the x-ray beam; this position is used for HRXRD analysis.

with no compensation for tilt, so it is the one normally used in production tools. The determination of tilt is illustrated in Figure 3.8.

This method has become very widely used. Virtually all compound semiconductor manufacturing lines use this measurement routinely for checking the settings of the growth reactors, especially after some change, e.g., of a precursor gas bottle or a break in a molecular beam epitaxy (MBE) system. It is used directly for composition measurement in ternary materials such as AlGaAs and InGaAs. In quaternary layers, one degree of freedom remains and a combination with another method, such as photoluminescence or XRF, is required. In silicon metrology, the method is used for the determination of composition (and thickness, as seen in Chapter 2) of the small selected-area epitaxial growth of SiGe. In the latter case, SiGe is always grown on symmetric (001) surfaces, so no tilt compensation is required.

3.5 Summary

- XRR gives limited information on composition of amorphous solids, mainly through determination of the electron density.
- GIXRF gives additional information and may be useful in measuring layers of different composition but similar electron density.
- XRD both identifies phases and measures their composition in polycrystalline films through the positions and intensities of peaks.

- Absolute lattice parameter measurements give composition measurement in bulk crystals (boules or wafers) if there is a single dopant/solute whose effect on the lattice parameter is known.
- Relative lattice parameter measurements (HRXRD) are commonly used to perform similar measurements in binary or ternary alloys of silicon or compound semiconductors.
- Because of the narrow, sharp peaks and the high angular resolution of HRXRD tools, the accuracy and repeatability of composition measurement in epilayers is very good, typically 0.5% accuracy and 0.1% repeatability.

References

1. D. Hausermann and M. Hart, *J. Appl. Cryst.*, 23 (1990) 63–69.
2. S.P. Tobin, J.P. Tower, P.W. Norton, D. Chandler-Horowitz, P.M. Amirtharaj, V.C. Lopes, W.M. Duncan, A.J. Syllaios, C.K. Ard, N.C. Giles, J. Lee, R. Balasubramanina, A.B. Bollong, T.W. Steiner, M.L.W. Thewalt, D.K. Bowen, and B.K. Tanner, *J. Electron. Mater.*, 24 (1995) 697–705.

4

Strain and Stress Metrology

4.1 Introduction

Measurements of the orientation and spacing of a crystal plane are the most fundamental x-ray diffraction measurement, stemming directly from the Bragg law ($n\lambda = 2d\sin\theta$). Differentiating the latter, we obtain

$$\frac{\delta d}{d} = -\delta\theta.\cot\theta \tag{4.1}$$

The higher the Bragg angle, θ, the greater the sensitivity to strain, $\delta d/d$. It is fairly straightforward to detect peak shifts (in the scattering angle, 2θ) in polycrystalline materials to 0.01° by x-ray diffraction (XRD). At an intermediate Bragg angle (45°) this corresponds to a strain of 200 ppm. In epitaxial structures one can do at least 10 times better by high-resolution x-ray diffraction (HRXRD). With special methods (not suitable for fabrication line tools) another order of magnitude sensitivity is possible. With topographic methods, changes in strain of this order can be mapped across the whole wafer.

To convert lattice spacing to strain we need the lattice spacing of the unstressed planes. This may be known (e.g., in single-crystal silicon) or may be measured from the spacing of planes parallel to the specimen surface. As long as there are no significant changes of strain within the penetration depth of the beam, this gives the undistorted spacing, since there can be no residual stress normal to the surface of a solid, at the surface itself. To convert this spacing to stress we need to know the Young's modulus of the material, noting that this may be different in a thin film from that in the bulk. Shear strains do not affect the interplanar spacing but only the unit cell symmetry. This cannot normally be measured in polycrystalline thin films, but can with some trouble be measured in single crystals.

Stress metrology in thin films is a key element in the development of a deposition process. The stresses can be extremely high and can vary substantially with factors such as the gas pressure in a sputtering deposition system. This affects the durability and adhesion of the films and their robustness to further processing.

Single-crystal wafers and epilayers are normally free from such gross strains. However, they often contain microcracks and dislocations and sometimes strain fields from other defects. These can easily prevent a device from working, or diminish its performance or service lifetime. It is therefore also necessary to have a defect inspection metrology for these much more subtle strains.

4.2 Strain and Stress in Polycrystalline Layers

4.2.1 $\sin^2\psi$ Analysis

The strains in a thin film are normally biaxial. That is, they consist of tensile or compressive strains parallel to the surface. If there is a gradient of such strains, then the strain must be triaxial, in which there is a component of strain normal to the surface. Conversely, a triaxial strain within the material necessarily involves gradients of strain normal to the surface, again because the normal strain at the surface must be zero.

We therefore want to make measurements of the change in lattice parameter parallel to the surface. Recently, it has become possible to measure this simply and directly, using grazing incidence in-plane x-ray diffraction (GIIXD), described below, but first we describe the conventional $\sin^2\psi$ method. This involves measuring the shift in peak positions of a single reflection for grains that are differently oriented with respect to the surface. The method is outlined in the inset box, and Figure 1.9 is repeated as Figure 4.1. It is necessary to take several measurements with the diffracting plane at different (ψ) angles to the surface, and to measure the slope of the lattice spacing of the diffraction plane as a function of $\sin^2\psi$. This is best achieved with a scan of the χ axis (refer to the axis definitions in the inset box on p. 8 over ±45°. Accuracy improves considerably as the Bragg angle is increased, from the behavior of the cotangent term in Equation 4.1. More advanced analysis, of triaxial strains and strain gradients, will be discussed in Chapter 10.

4.2.2 GIIXD Analysis

The recent use of GIIXD for stress measurement is a very useful development for the films used in the semiconductor industry. The advantages are:

1. The measurement direction is almost parallel to the crystal surface, so the biaxial strains are measured directly.

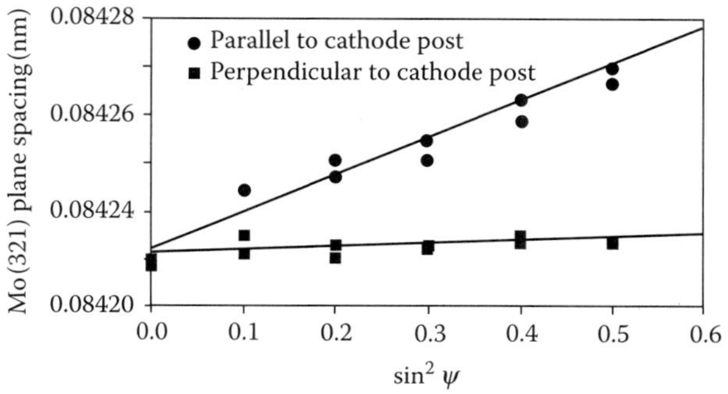

FIGURE 4.1
Plot of (321) lattice plane spacing in Mo (measured from the shift of the diffraction peak) as a function of exit angle (), plotted using the function sin^2. The sample is magnetron-sputtered Mo on glass. The biaxial surface stresses are found from the slopes of the two lines. In this example the stress is anisotropic, with different values parallel (481 ± 56 MPa) and perpendicular (99 ± 56 MPa) to the cathode post. (From B.L. Ballard et al., *Advances in X-Ray Analysis*, 37, J.V. Gilfrich et al., Eds., Plenum Press, New York, 1994. With permission.)

2. Though the signal is low in this geometry, the noise is greatly reduced, so signal/noise ratios are improved.
3. Control of the grazing incidence angle may be used to control the depth penetration, and thus reveal stress gradients.
4. The method may be used on a textured sample to give sharp high-angle reflections. With the conventional method, only one reflection is available, corresponding to the texture of planes parallel to the surface. This reflection may be relatively low angle, and it is usually broadened by the film thickness.

In most laboratory-based systems, the divergence of the focused x-ray beam is relatively large and there is little depth sensitivity. However, the average penetration of the grazing incidence x-ray beam is small and there is excellent surface sensitivity, making it highly appropriate for measurements on thin films of metals such as Ru, Pt, and Cu. An example of a GIIXD scan from a Cu thin film, taken with a microfocus source with focusing optics, is shown in Figure 4.2. The peak positions index to the face-centered cubic structure with lattice parameter equal to that of bulk copper, so in this case the film is unstressed.

Uniform strain results in a shift of the Bragg peak position, whereas strain dispersion results in peak broadening. Analogously to Equation 4.1, the broadening of the peak is related to the strain dispersion $\Delta\varepsilon$ by

$$\Delta(2\theta)\cos\theta = 2\Delta\varepsilon\sin\theta \qquad (4.2)$$

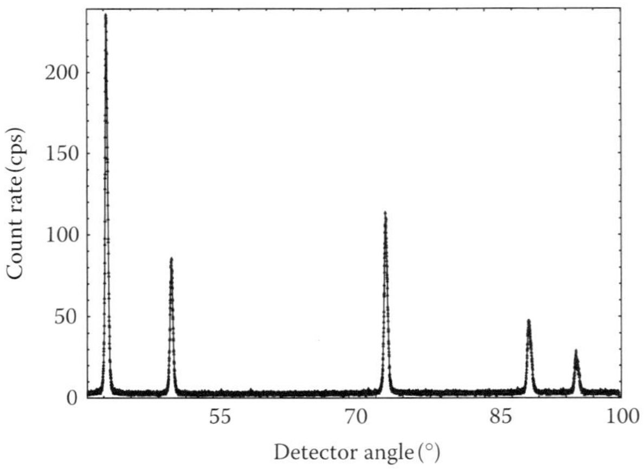

FIGURE 4.2
GIIXD scan of the scatter from a thin film of Cu on Si, taken with an x-ray beam from a Bede Microsource® generator and a low-divergence polycapillary optic. A Ni foil removed the CuKβ line, and a combination of Soller slits and graphite analyzer set the instrumental resolution at about 0.15° with low background.

As explained in Chapter 5, the peak width is also influenced by the grain size, and as a result, the total peak width is given by

$$\Delta(2\theta)\cos\theta = 2\Delta\varepsilon\sin\theta + \lambda/L \tag{4.3}$$

where $\Delta\varepsilon$ is the strain dispersion, L is the grain size in the direction normal to the Bragg planes, and λ is again the x-ray wavelength. When the full-width-at-half-height maximum, $\Delta(2\theta)$, of the peaks in Figure 4.2 is plotted on a Williamson–Hall plot (Section 1.4), that is, $\Delta(2\theta)\cos\theta$ vs. $\sin\theta$, we find that the gradient is almost zero. From this we conclude that there is also almost no strain dispersion in the film. From the intercept at $\theta = 0$, we find that the average in-plane grain size is 33 ± 2 nm.

There is a large body of literature concerning attempts to use the Bragg peak shape to determine the internal state of the material. This is not a reliable metrological technique. As a rough guide to interpretation we can say that:

1. Broadening of peaks resulting in a Gaussian peak shape indicates that the cause of broadening is strain.
2. Broadening of peaks resulting in a Lorentzian peak shape indicates that the cause of broadening is a small grain size.

Most peaks, perhaps unsurprisingly, are best fitted by pseudo-Voigt functions. These consist of a mixture of Gaussian and Lorentzian components.

In the thin film example given, Figure 4.3, the gradient was almost zero, indicating almost no strain dispersion present. This is not always the case,

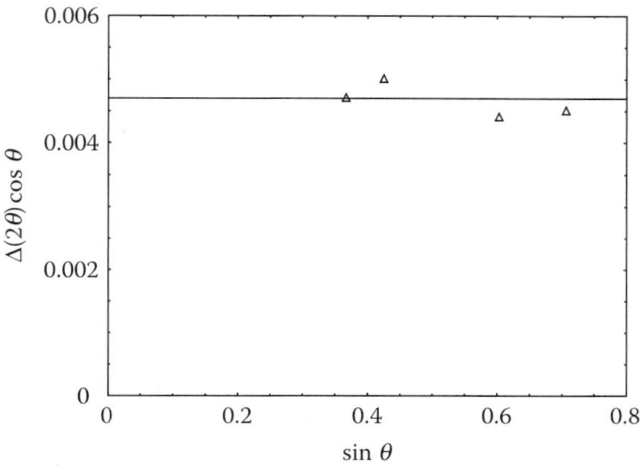

FIGURE 4.3
Williamson–Hall plot of GIIXD data from Figure 4.2.

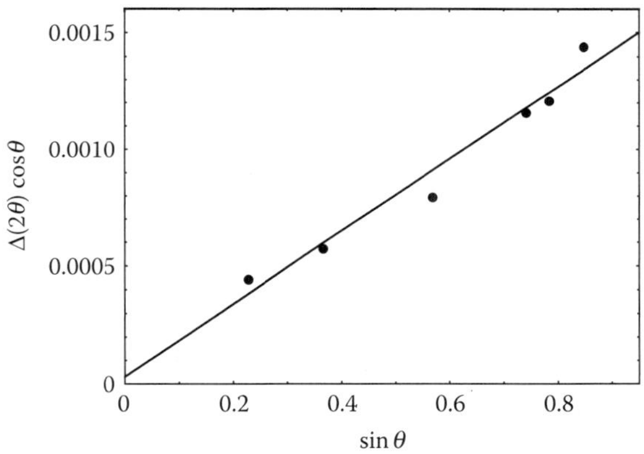

FIGURE 4.4
Plot of the product of the diffraction peak broadening and $\cos\theta$ against $\sin\theta$ (Williamson–Hall plot). The material is an annealed Al_2O_3/SiC nanocomposite and the data taken with x-rays of wavelength 1.55 Å. The grain size is given by the intercept at $\sin\theta = 0$ and the microstress by the slope.

as for the Al_2O_3 nanocomposite shown in Figure 4.4. Here, the intercept of the Williamson–Hall plot enables us to determine the grain size to be large, at 5 μm. The fitting procedure gave a rather poor precision, +9 and –5 nm, and this arises from the small value of the intercept. Note that the precision of measurement of the microstrain dispersion, derived from the gradient of the curve, is much better. Here we find the strain dispersion to be $7.8\ (\pm 0.8) \times 10^{-4}$.

With a low divergence beam in the direction normal to the specimen surface, i.e., almost normal to the scattering plane, the GIIXD method can be used as a probe of the composition, lattice parameter, and lattice parameter dispersion as a function of depth. An example, in this case taken at the

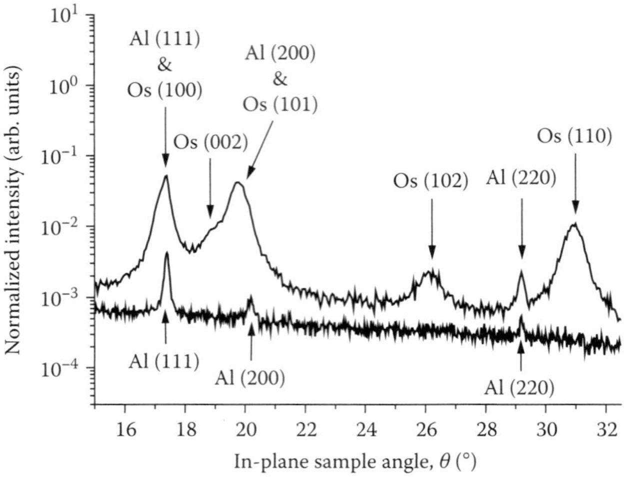

FIGURE 4.5
GIIXD profile of a 5-nm Os, 5-nm Al bilayer sputtered on thermally oxidized silicon. The lower curve is taken at an incidence angle of 0.2°, the upper at 0.55°. (Reused with permission from J.D.R. Buchanan, *Journal of Applied Physics*, 96, 7278 [2004]. Copyright 2004, American Institute of Physics.)

European Synchrotron Radiation Facility in Grenoble, is of a bilayer of 5 nm Os and 5 nm Al sputtered onto a thermally oxidized Si substrate. At very low incidence angle, below the critical angle of Al, the evanescent x-ray wave does not penetrate sufficiently to reach the underlying Os layer, and thus only the Al peaks are visible in the GIIXD profile (Figure 4.5). For a larger incidence angle, the wave penetrates to the Os layer and we see peaks from both Os and Al layers.[1]

The integrated intensity under the diffraction peaks can be modeled from the depth penetration. Figure 4.6 shows the integrated intensity as a function of incidence angle for the W(110) diffraction peaks from a similar 5-nm W and 5-nm Al bilayer. The data points are the experimental measurements, with the solid line simulated assuming that the scattered amplitude is proportional to the strength of the electric field as a function of depth. The dashed line shows the calculated depth at which the incident electric field falls to 1/*e* of its surface value. The maximum intensity corresponds to the angle at which the wave just penetrates into the W layer, the subsequent slow fall being associated with increased absorption as the incidence angle is increased.

While interplanar spacing is the fundamental measured parameter, and can be converted to strain with little uncertainty, stress is the parameter needed by a thin-film engineer. It is stress that affects film adhesion and delamination. The conversion of strain (ε) to stress (σ) is performed by the well-known relationship, $\sigma = \varepsilon E$, where E is the Young's modulus or, more properly, the full tensor expression. For films thicker than, say, 10 nm, it is safe to take the bulk value of modulus as long as the film is in the same structural state as the bulk (for example, one should not use crystalline parameters for an amorphous film).

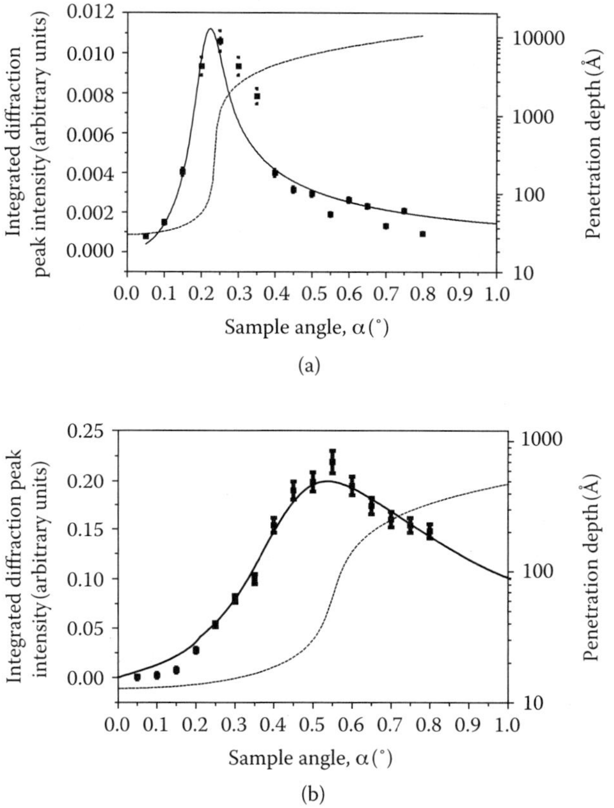

FIGURE 4.6
Integrated intensity of (a) the Al (111) and (b) the W (110) GIIXD diffraction peak from a 5-nm W, 5-nm Al bilayer as a function of incidence angle. The solid curves are simulations of the diffracted intensity; the dashed curves represent the penetration depth of the wave as a function of incidence angle.[1] (Reused with permission from J.D.R. Buchanan, *Journal of Applied Physics*, 96, 7278 [2004]. Copyright 2004, American Institute of Physics.)

However, it is not at all obvious what parameters one should use for extremely thin films, which may comprise crystalline domains in an amorphous, oxide, or graded-density matrix (see, e.g., Welzel et al.[2]). It may be possible to use information from such methods as the nanoprobe or ultrasonic AFM to validate the conversion for such films. Alternatively, the constants may be obtained from calibration against another method, such as bowing of a thin substrate.

4.3 Relaxation of Epitaxial Layers

If the interface between a strained layer and a single crystalline substrate is only partially coherent, i.e., it contains interface misfit dislocations, it is said

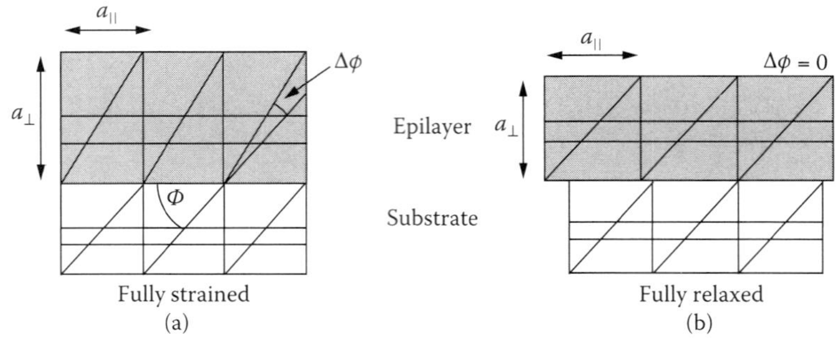

FIGURE 4.7
A side view of (a) coherent and (b) partially relaxed epilayers. The relaxation process changes both the interplanar spacings of the epilayer and the angles between the reflecting planes and the surface.

to be relaxed. Figure 4.7 shows a coherent and a relaxed layer, and it is clear that both the mismatch and the orientation of the asymmetric planes change between the substrate and the layer. The *tetragonal distortion* changes as soon as the layer starts to relax.

Layer relaxation may be either desirable or undesirable. It is desirable in, for example, the growth of relaxed layers of SiGe to form a substrate on which a thin layer of strained silicon can be grown coherently. Even if the layer is grown coherently, it may relax on further processing if the layer thickness is larger than the critical thickness and the layer is thus in a metastable state. Relaxation is generally undesirable if the SiGe is being used for a transistor element, since changed strain state and carrier diffusion along the dislocation lines will alter the electric properties locally and probably destroy the transistor action. In either case, a metrology method is required to determine the strain state and existence of these detrimental misfit dislocations.

We need to measure the lattice parameter both parallel and perpendicular to the interface, and must eliminate the effect of the tilt of the layer relative to the substrate. Tilt means that the growth direction, nominally (001) for most Si wafers, differs for the substrate and the epilayer. The out-of-plane lattice parameter can be measured easily using a symmetric reflection, like (004). We then need an additional measurement with a component of the diffraction vector parallel to the surface. This means an asymmetric reflection from a plane that is at as high an angle to the surface as possible. 224, 311, and 511 are all acceptable.

Now, the effect of tilt on the splitting is reversed if the specimen is rotated by 180° about its surface normal, but the splitting due to the mismatch will not be affected by such a rotation. Thus, we may make grazing incidence or grazing exit measurements (Figure 4.8) to separate the tilt from the true splitting.

The mathematics of calculating the relaxation is simple if three reflections (symmetrical, grazing incidence, and grazing exit) are used. However, it is

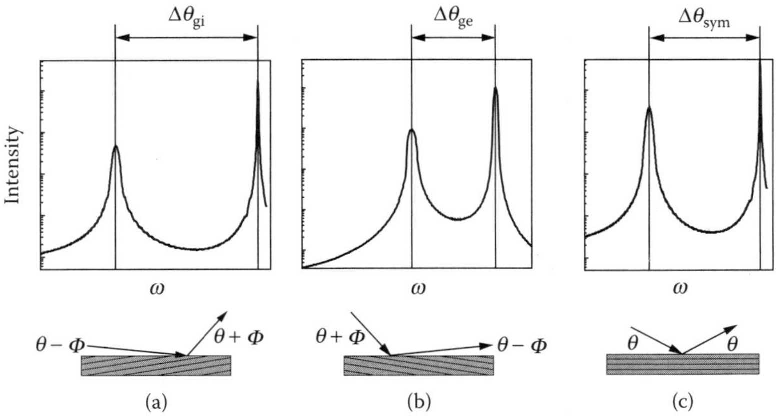

FIGURE 4.8
(a) Grazing incidence geometry. (b) Grazing exit geometry.

necessary to have a strong enough layer signal to measure the peak in a rocking curve with an open detector, since the angle of scattering is in principle unknown. Also, this method restricts throughput, since in principle, only two measurements are needed for the two unknowns. Though the analytic derivation of relaxation for two arbitrary reflections is complex, it is straightforward to solve by numerical iteration. This approach is used exclusively in fab tools.

The problem of thin layers is essentially one of signal/noise. A relaxed peak is broadened by the dislocation content of the layer. Hence, it is difficult to distinguish from the background, which will itself be increased from the diffuse scatter from the dislocations. But a detector slit will reduce the diffuse scatter and improve signal/noise by up to 50 times. This makes it possible to measure relaxation in layers as thin as 20 nm.

However, we then have a classic dilemma. The slit on the detector greatly restricts its angular acceptance. We need to measure either the ω or the 2θ angle for the relaxed peak. But we do not know where to place the detector unless we know the relaxation. Until recently, this required a two-dimensional search for the relaxed peak, which could take several hours. However, a new method has been invented by Matney et al.[3] A symmetrical 004 scan is taken, which measures the lattice parameter perpendicular to the surface. From this we may calculate the locus of a line in $\omega - 2\theta$ space on which the asymmetric peak must lie. The principle is that if the layer is fully relaxed, the orientations of the crystal planes in substrate and layer must be the same in symmetric and asymmetric reflections; if the layer is fully strained, the lattice parameter parallel to the surface must be the same in both cases. From these conditions, the locus of the line, which we call the relaxation axis, can be calculated. This is most easily seen in reciprocal space and will be shown in detail in Chapter 11.

The relaxation axis is of finite length and extends from 0 to 100% relaxation. Thus, provided the diffractometer axes can be driven with arbitrary coupling, a simple scan along the relaxation axis measures the degree of relaxation.

4.3.1 Relaxation in SiGe

An example of relaxation scans for SiGe was shown in Figure 1.12. The peaks for the two relaxations scans shown are different. The unrelaxed example is quite sharp, whereas the other is much broader. It is also not uncommon to see irregular and asymmetric peaks in relaxation scans. These will arise from differing degrees of local relaxation in regions covered by the x-ray beam. If a strained silicon layer is grown on a SiGe virtual substrate, the carrier mobility will vary slightly from point to point. Although we report relaxation as a single number, the device engineer does have more detailed information available from relaxation scans if device behavior appears to be inconsistent. In most cases, the relaxation can be determined to a repeatability of better than 1%.

The earliest onset of relaxation is best determined by the diffraction imaging method introduced in Chapter 1 and discussed in detail in Chapter 12. Figure 4.9 shows misfit dislocations in a p/p^{++} silicon sample with a doped homoepilayer. They have penetrated from the wafer edge on annealing, and this low degree of relaxation would be very hard to find on a diffractometer measurement. The inhomogeneity of the distribution is also clear.

In many cases, SiGe is grown on Si only in selected areas, not as a blanket film. We discuss the metrology of such areas in Chapter 12, and Figure 12.18

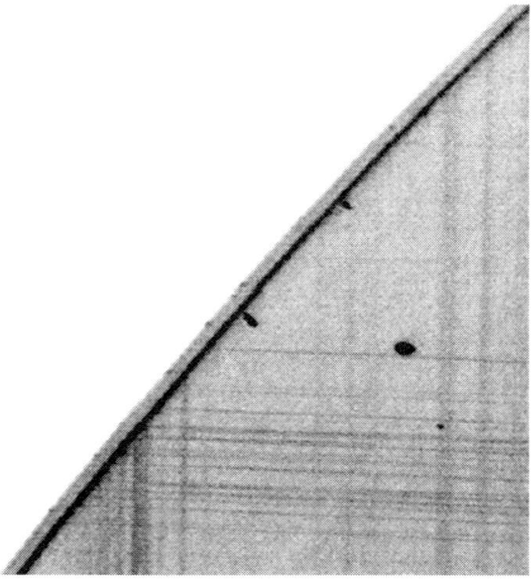

FIGURE 4.9
XRDI (BedeScan™ image) of misfit dislocations in p/p^{++} epitaxial doped Si (white). MoKα, 220 in transmission. The wafer edge is at the upper left-hand side of the image.

shows how ray diffraction imaging (XRDI) can be used to detect the onset of relaxation in selected areas only 300 × 300 μm^2 in size.

4.3.2 Relaxation in Compound Semiconductors

Relaxation has been a particular problem in compound semiconductors because the mismatch in the plane of the film between the two relaxed lattices of different lattice plane spacing is accommodated by misfit dislocations lying in the interface. These misfit dislocations, which can be revealed by a number of techniques, including transmission electron microscopy, x-ray diffraction imaging, and atomic force microscopy, act as electron traps. There is localized heating, which results in rapid degradation and failure of lasers and light-emitting diodes made from III-V semiconductors. (For reasons that are still not fully understood, misfit dislocations do not have such a deleterious effect on nitride-based light-emitting devices.)

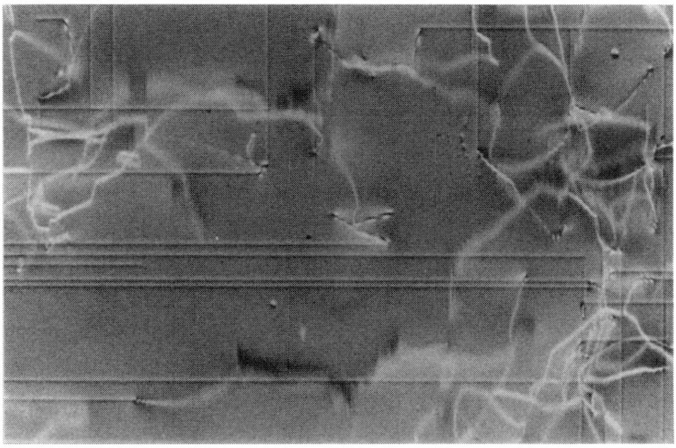

FIGURE 4.10
Transmission x-ray diffraction image of a thin InGaAs layer on GaAs in the very early stages of relaxation. The straight lines are the 60° type misfit dislocations, and the curved lines the associated threading dislocations penetrating the layer from the substrate.

The relaxation process starts from threading dislocations that are driven across the wafer by the biaxial stress in the film, trailing the straight misfit dislocation in the interface. Figure 4.10 shows both the threading and misfit dislocations in an x-ray topograph of an $In_xGa_{1-x}As$ layer on GaAs in its early stage of relaxation. In the very early stages of relaxation, the displacement of the Bragg peak of the layer with respect to the substrate is too small to measure with high-resolution x-ray diffraction. The onset of relaxation is therefore quite difficult to determine and depends critically on the sensitivity of the technique used. High-resolution x-ray diffraction is much more sensitive than Raman spectroscopy or transmission electron microscopy; diffraction imaging is the most sensitive of all the techniques as individual misfit dislocations are imaged. Using *in situ* imaging during MBE growth at the European Synchrotron Radiation Facility in Grenoble, Parbrook et al.[4] were able to follow the relaxation as $In_{1-x}Ga_xAs$ grew on GaAs and identify the thickness at which the very first misfit dislocations nucleated. They measured the critical thickness for both nucleation and multiplication of misfit dislocations in Si-doped material and were able to fit the data to theoretical predictions based on a force balance model.

When substantial relaxation occurs, the misfit dislocation distribution is nonuniform, and this results in a broadening of the x-ray rocking curve peaks from the layer. Use of a very small x-ray spot reduces this broadening; measurement with a small spot as a function of position on the wafer reveals the nonuniformity.

A feature of relaxation in compound semiconductors is that it is often asymmetric; that is, the relaxation is not independent of direction in the film plane. If the lattice parameter difference between substrate and film in an (001) film is measured using the surface symmetric 004 reflection and the asymmetric $2\bar{2}4$ reflection, the relaxation in the $1\bar{1}0$ direction can be determined. This will often be different from that measured in the $[\bar{1}10]$ direction using the $\bar{2}24$ reflection. The origin of the asymmetry lies in the different dislocation velocities on the fast and slow glide planes of the zinc-blende structure. Tanner et al.[5] observed that the first misfit dislocations to nucleate and multiply were always of the fast type.

4.3.3 Relaxation in Compounds Based on GaN

The GaN-based semiconductors, such as $Al_xGa_{1-x}N$, have no substrates with near-matched lattice parameters. Consequently, an almost fully relaxed buffer layer of GaN is grown on the sapphire substrate prior to growth of the active epilayers that form part of the devices. With the high dislocation densities involved, peaks tend to be broad, which reduces the precision of the measurement of relaxation.

A more direct approach to measurement of relaxation is to determine the in-plane lattice parameter difference between film and substrate directly using grazing incidence in-plane diffraction. Here, an analyzer is used in exactly the same way as for standard high-resolution x-ray diffraction to separate tilt and dilation. An example of the combination of out-of-plane and in-plane diffraction is the work of Lafford et al.,[6] who studied the relaxation in $Al_xGa_{1-x}N$/AlN/GaN epilayers grown on sapphire using triple-axis diffraction. From the analyzer peak position in the symmetric 00.2 and in-plane diffraction 11.0 reflections, the *a* and *c* lattice parameters as a function of the $Al_xGa_{1-x}N$ layer thickness (Figure 4.11) were determined independently of the Al fraction. As the AlN interlayer thickness increased, the in-plane lattice parameter of the $Al_xGa_{1-x}N$ layer decreased. It therefore became more strained with respect to the underlying AlN interlayer as the AlN interlayer thickness increased.

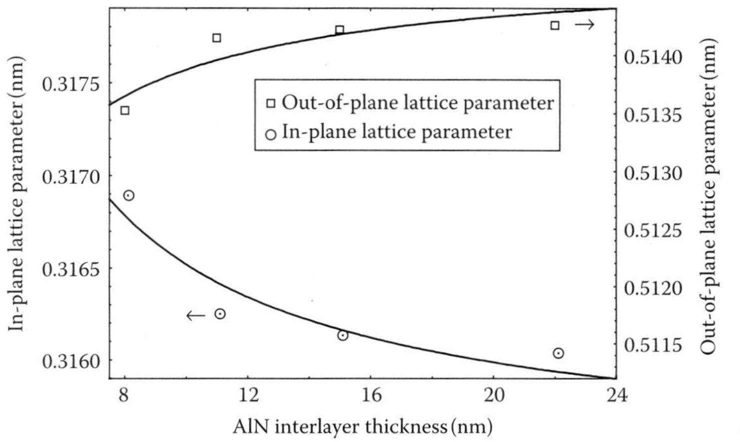

FIGURE 4.11
In-plane and out-of-plane lattice parameters of a thick overlayer of $Al_xGa_{1-x}N$ as a function of AlN interlayer thickness.[6] (Reused with permission from T.A. Lafford, *Applied Physics Letters*, 83, 5434 (2003). Copyright 2003, American Institute of Physics.)

4.4 Thin Strained Silicon Layers

Significant advances in transistor switching speeds (between 10 and 40%) can be achieved by use of silicon that is strained in its plane. As a result of associated changes in the electronic band structure, tensile strain improves *n*-type conductivity and compressive strain improves *p*-type material. Methods of achieving the strain include depositing the silicon on a SiGe layer of different lattice parameter, surrounding the layer with trenches filled with SiGe, and overcoating the whole device at high temperature with a film having a different coefficient of thermal expansion from that of the layer.

The strain is thus a critical part of the device performance and a suitable metrology is needed. Layers are normally just thick enough (typically 10 to 20 nm) to be measured by both HRXRD and XRR. However, if it is suspected that they may have relaxed in the processing, they should be measurable separately from their Si or SiGe substrate. The problem is that the diffracted signal from the substrate has long tails and adds to the weak signal from the strained layer. An example of a double-crystal HRXRD rocking curve of a nominally 16-nm strained Si layer on a virtual substrate of 1 μm of SiGe linearly graded between zero and 15% Ge, a 67-nm layer of $Si_{0.85}Ge_{0.15}$, and an 8-nm layer of $Si_{0.7}Ge_{0.3}$ is shown in Figure 4.12. The broad peak to the left arises from the SiGe; a triple-axis measurement reveals a sharp spike at the very left of the plot due to the 67-nm $Si_{0.85}Ge_{0.15}$ of constant composition. (Wafer curvature and the effect of mosaic structure results in the peak being smeared out of the double-crystal plot shown.) The very intense sharp peak is from the Si substrate, while the small peak on the far right arises from the strained Si. We note that the lattice parameter is lower than that of the Si substrate, as the crystal planes measured are parallel to the surface. The Si is strained to be lattice matched with the top SiGe, and hence there is a tensile strain in the film plane, as SiGe has a larger lattice parameter than Si. As the volume of the Si unit cell must remain constant, the lattice contracts in the out-of-plane direction. Hence the 004 Bragg peak of the strained Si is at a higher value than that of the Si substrate.

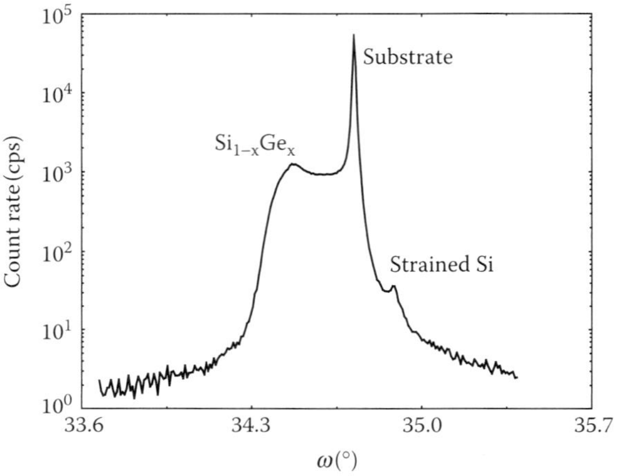

FIGURE 4.12
Double-axis HRXRD rocking curve of a strained (001) Si structure. Symmetric 004 reflection. The broadening of the feature associated with the $Si_{1-x}Ge_x$ layer arises through variation of x with depth. (Courtesy S.R. Evans, M.Sc. thesis, now at University of Edinburgh.)

We note that Figure 4.12 is on a logarithmic scale and that the strained Si peak is weak compared with that from the substrate. If significant relaxation occurs, then the strained Si peak moves to the left and rapidly becomes indistinguishable from the substrate signal. This can be overcome by choosing the GIIXD geometry, where, as we have seen, the x-ray wave does not penetrate deep into the material. Thus, there is virtually no signal from the Si substrate. There is an additional advantage to the GIIXD geometry in that the lattice parameter is measured in the plane of the film, and hence the relaxation determined directly without further computation or knowledge of the Poisson ratio. Unfortunately, due to the weak signal, the x-ray beam intensity must be high, and correspondingly there will be a somewhat low resolution. The sensitivity to relaxation in the GIIXD measurement is thus less than in out-of-plane HRXRD measurements.

4.5 Whole Wafer Defect Metrology

Image contrast in x-ray diffraction imaging (XRDI) is almost entirely governed by the strain in the crystal lattice. As a large-area imaging technique, it is ideally suited for the mapping of strain-related defects across a whole wafer. Recent advances in instrumentation have resulted in tools that are automated, completely digital, and fast enough to use in a fab line. A complete wafer can be scanned in about 30 min at medium resolution, and suspect areas can be (automatically) examined at greater resolution.

The transmission image in Figure 1.13 has already shown the image of a whole wafer, in which growth dislocations are seen at the center, and thermal slip dislocations formed in slip bands from defects. Dislocations formed at interfaces in heteroepitaxy are shown in Figure 4.9. Regions of high defect density and their locations are often needed, to correlate with yield losses. These can be produced automatically by standard image processing methods (see Figure 12.14 and the associated discussion). The x-ray image can also be correlated with and superimposed on an optical micrograph. Relaxation may be studied directly by XRDI as discussed in Section 4.3.1.

Polishing and edge-shaping damage is an important problem in substrate preparation, since defects introduced may carry through into the grown layers. In our experience the most common problem is edge-shaping damage, as shown in Figure 4.13. The tiny defects revealed around the wafer edge are likely to grow into the long slip bands (such as those seen in Figure 1.13, although these were from different types of nucleation) under processing such as rapid thermal annealing (RTA).

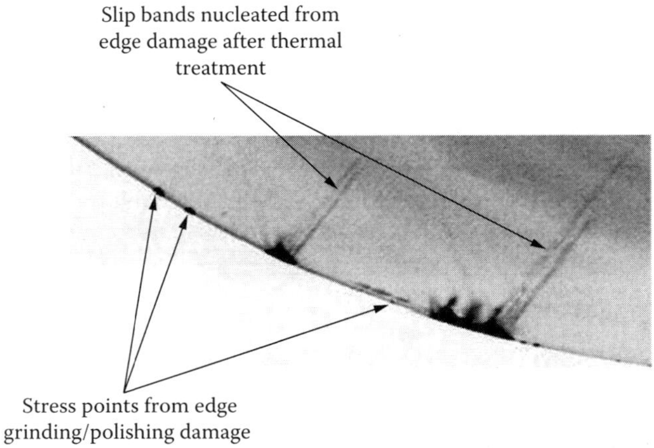

FIGURE 4.13
Si wafer, transmission 004 reflection, showing edge defects (microcracks), some of which have grown into slip bands on subsequent thermal treatment. (Courtesy Dr. DonKun Lee, LG Siltron, Korea.)

It should be noted that this kind of strain mapping is quantitative. The digital data acquisition records the small orientation changes at each part of the specimen. Data may be processed so as to enhance or minimize the visibility of long-range strain fields and curvatures, and the orientation changes themselves may be mapped over the wafer surface. A figure of merit may be generated to provide a pass/fail criterion for an individual wafer.

In very thin crystals, less than about a third of a parameter known as the extinction distance (inversely related to the strength of the Bragg reflection excited and typically several tens of micrometers), dislocation images become invisible unless high-resolution x-ray optics are used. However, this is not the case for thin, submicrometer films on wafers. The strain fields of the dislocations are long range, and even if the dislocation is in the film or at the interface, the strain field extends deep into the substrate. Imaging the topograph with the (strong) substrate reflection then reveals the dislocation. Nevertheless, for misfit dislocations at interfaces in very thin epitaxial layers, the effect of the surface is to cancel the long-range strain field. Misfit dislocations at the interface of nanometer-thick films do not image in x-ray topographs.

4.6 Summary

- X-rays are an extremely sensitive method of measuring strain. This applies to polycrystals and single crystals and to films 3 nm thick.
- In polycrystalline films, strains may be measured by conventional XRD or by the newer grazing incidence in-plane diffraction (GIIXD), to obtain biaxial strains and to indicate triaxial strains or strain gradients.
- With knowledge of the Young's modulus, strain may be converted to stress, an important parameter in multilayer growth processes. However, the Young's modulus values for extremely thin layers may not be well known.
- Relaxation in epitaxial layers (including selected area epitaxy in a 100-μm window) can be measured accurately and rapidly.
- The strain in strained silicon layers can be measured by means of GIIXRD. If the layer is isolated from the substrate, then strain may be measured in a small spot.
- X-ray diffraction imaging (XRDI) shows strain fields as an image akin to electron microscopy but on a wafer scale.

- Whole-wafer XRDI shows strain fields caused by growth and slip dislocations, polishing damage artifacts, and precipitates. These include damage caused in thermal processing.
- Edge scan XRDI reveals defects from polishing damage that are likely to develop into slip bands during RTA.

References

1. J.D.R. Buchanan, T.P.A. Hase, B.K.Tanner, C.J. Powell, and W.F. Egelhoff, Jr., *J. Appl. Phys.*, 96 (2004) 7278–7282.
2. U. Welzel, J. Ligot, P. Lamparter, A.C. Vermeulen, and E.J. Mittemeijer, *J. Appl. Cryst.*, 38 (2005) 1–29.
3. K.M. Matney, P.A. Ryan, and P. Feichtinger, A Fast, Direct and Automated Measurement of Layer Relaxation Using High-Resolution X-Ray Diffraction, paper presented at the EMC Conference, Notre, Dame, 2004.
4. P.J. Parbrook, B.K. Tanner, B. Lunn, J.H.C. Hogg, A.M. Keir, and A.D. Johnson, *Appl. Phys. Lett.*, 81 (2002) 2773–2775.
5. B.K. Tanner, P.J. Parbrook, C.R. Whitehouse, A.M. Keir, A.D. Johnson, J. Jones, D. Wallis, L.M. Smith, B. Lunn, and J.H.C. Hogg, *Appl. Phys. Lett.*, 77 (2000) 2156–2158.
6. T.A. Lafford, P.J. Parbrook, and B.K. Tanner, *Appl. Phys. Lett.*, 83 (2003) 5434–5435.

5

Mosaic Metrology

5.1 Grain Size Measurement

For diffraction from polycrystalline materials, in which each grain scatters independently and is not too thick (above a few micrometers), intensities and shapes of diffraction lines are given adequately by the kinematic theory, discussed in Chapter 10. An important result of this theory is that the angular width of a diffraction line is inversely proportional to the number, N, of unit cells along the diffraction vector (the normal to the diffracting planes). Thus, the broadening is inversely proportional to the grain size.

In a thin film, the grain size perpendicular to the surface is at maximum the thickness of the film. Thus, diffraction lines are often very broad. This is especially true in the Bragg–Brentano method, where the diffracting vector is always normal to the surface. In the parallel beam methods, especially the new GIIXD method, the apparent thickness can be very large and broadening from this cause eliminated.

The line broadening may be used as a metrology for grain size. The Scherrer equation, derived from the kinematic theory, gives the grain diameter, t, in terms of the line broadening, $\Delta(2\theta)$ (in radians), as

$$t = \frac{0.9\lambda}{\Delta(2\theta)\cos\theta_B} \tag{5.1}$$

or

$$\Delta(2\theta)\cos\theta_B = \frac{0.9\lambda}{t} \tag{5.2}$$

where λ is the wavelength and θ_B the Bragg angle. The factor of 0.9 arises from the assumed line shape of the diffraction peak and the assumed crystallite shape, and is only approximate. In the Williamson–Hall plot it is set at one. Nevertheless, Equation 5.1 does provide a metrological tool, albeit of limited absolute accuracy, by allowing comparison between samples measured and analyzed in the same way. However, grain size metrology is confused by broadening from two other causes:

- Instrument broadening, due to beam divergence, slit sizes, etc.
- Microstrains in the material

These must be separated. Instrument broadening is measured for a particular diffractometer and its configuration (slit sizes, etc.) by taking a measurement with a suitable strain-free and large-grain standard, such as a NIST LaB_6 standard reference material (SRM). The broadening must be subtracted from the measured full-width-at-half-height maximum (FWHM) to obtain $\Delta(2\theta)$. Again, the exact Bragg peak shape must be known to perform this deconvolution properly, but to a very good approximation, this subtraction can be done in quadrature. That is,

$$[\Delta(2\theta)]^2 = [\Delta(2\theta)_{measured}]^2 - [\Delta(2\alpha)]^2 \tag{5.3}$$

where $\Delta(2\alpha)$ is the instrument resolution function.

Separation of strain broadening from crystallite size broadening is possible because of the different dependencies on Bragg angle or crystal reflection. Another way of saying this is that the reflection width determined by the crystallite size is of constant width in a reciprocal space plot. Strain broadening, however, varies with Bragg angle, as we saw in Chapter 4. Taking Equation 4.1, rearranging and multiplying by 2 so that both tensile and compressive stresses are taken into account, we have

$$\Delta(2\theta)\cos\theta = 2\frac{\delta d}{d}\sin\theta = 2\Delta\varepsilon\sin\theta \tag{5.4}$$

Comparing Equations 5.3 and 5.4, we see that for a stress-free material the product of the peak width and the cosine of the Bragg angle remains constant. For a strained material, the product will have a slope that is positive and linear with $\sin\theta_B$. Thus, the two contributions may be distinguished and measured separately. A thin-film example of this Williamson–Hall plot of $\Delta(2\theta)\cos\theta$ vs. $\sin\theta$ is shown in Figure 5.1. (Note that the plot is often made against $(\sin\theta)/\lambda$ to allow measurements of grain size at different wavelengths to be compared directly.) The example is from grazing incidence in-plane x-ray diffraction (GIIXD) measurements on a thin film of TiN grown on Si, and the high value of the intercept compared with the examples in Chapter 4 shows immediately that this material is nanocrystalline. Even though the data are spread such that it is difficult to determine precisely the microstrain, the in-plane grain size can be determined with considerable confidence. From the straight-line fit, we can conclude that the mean grain size L is 7 (+2.5/–1.5) nm.

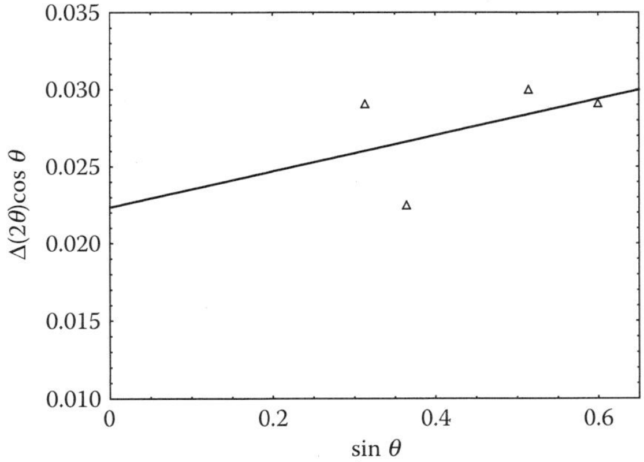

FIGURE 5.1
Williamson–Hall plot of the FWHM of the GIIXD Bragg peaks in a thin film of TiN on Si.

A complication — or opportunity — arises if the grains are not equiaxed. For platelets, for example, the peaks from reflections parallel to the platelet surface will be broader than those from oblique reflections. A nonmonotonic change of $\Delta(2\theta)$ with θ will then be expected. This may occur in thin films if the grain size within the film plane is larger than the film thickness and the grains have a preferred orientation (texture). Use of the GIIXD method often avoids this problem, as the texturing of thin films is often normal to the film plane (e.g., (111) texture in sputtered Cu).

5.2 Mosaic Structure in Substrate Wafers

Due to the huge effort expended on development over the past half century, the quality of silicon is now such that single crystals up to 450 mm diameter and over 1 m long can be grown with zero dislocation density throughout most of their length. The mosaic spread in wafers cut from such boules is zero. However, there are other materials of major importance to the microelectronics industry where this is not the case. Examples include GaAs and SiC. Here, the thermal conditions during growth result in the presence of local internal strains that are relieved by the creation of dislocations. When conditions are such that significant dislocation climb occurs, it is energetically favorable for the dislocations to form into low-angle boundaries, containing a high density of dislocations, separating regions of material with low dislocation density. These low-angle boundaries themselves have little long-range strain, and the resulting crystal has little internal strain dispersion, but a distribution of subgrains separated by the low-angle boundaries.

The mosaic spread within materials such as GaAs is now so low that it requires HRXRD in the triple-axis mode to measure the mosaic width. A high-resolution beam conditioner, usually a four-reflection monochromator, and a high-resolution Si or Ge analyzer are needed. The mosaic is measured directly from a rocking curve in which the analyzer is kept fixed and the specimen only is scanned.

In less perfect materials, the double-axis rocking curve will be significantly broadened, but always some form of analyzer, be it a slit or crystal, is needed to separate the mosaic from the strain dispersion.

5.3 Mosaic Structure in Epilayers

The most common origin of mosaic structure in epilayers lies in the relaxation process that occurs when the critical thickness is exceeded. In epilayers that are highly mismatched to the substrate, relaxation occurs at very low epilayer thickness and the density of misfit dislocations at the interface is very high. It is energetically favorable for the misfit dislocations and the associated threading dislocation segments to adopt nonuniform configurations. In the case of cube-oriented material with the diamond or zinc-blende structure (e.g., Si or GaAs, respectively), the misfit dislocations are of 60° type and have a mixture of screw and edge character. Bundles of misfit dislocations result in regions of rapid tilt-separating regions where the misfit dislocation density is low, giving rise to a mosaic structure out of the plane of the film. As with substrate wafers, the mosaic distribution can be separated from the strain distribution by recording a rocking curve in the triple-axis HRXRD geometry. An example of data from a thick epilayer of $Hg_{1-x}Mn_xTe$ grown by organometallic vapor phase epitaxy on (001) GaAs is shown in Figure 5.2. The solid line is a coupled specimen detector ($\omega - 2\theta$ scan) that measures the strain dispersion in the epilayer. This is narrow (12.5 arc sec) and is only slightly broadened from the instrument resolution. Thus, although the lattice mismatch between the epilayer and substrate is large, there is very little strain present in the almost fully relaxed film. The data points represent a rocking curve, that is, a scan of the sample at fixed detector position. This scan measures the mosaic width, which is found to be 170 arc sec. Characteristic of a random distribution of tilts, the mosaic distribution is quite well fitted by a Gaussian function (dashed line).

The evidence of Figure 5.2 suggests that the predominant cause of broadening of rocking curves in the double-axis HRXRD setting is from mosaic. This is often, but not always, the case. In Figure 5.3, we show an example of the 004 and 224 double-axis rocking curves from a 1-μm-thick layer of $In_{0.04}Ga_{0.96}As$, doped with 5×10^{18} cm^{-3} Si, grown by molecular beam epitaxy on (001) GaAs. Note that most of the broadening of the experimental rocking curves, compared with the width of the simulations, is due to the wafer curvature. This arises from the wafer and epilayer behaving like a bimetal strip and taking up the strain present in the epilayer by bending. Because

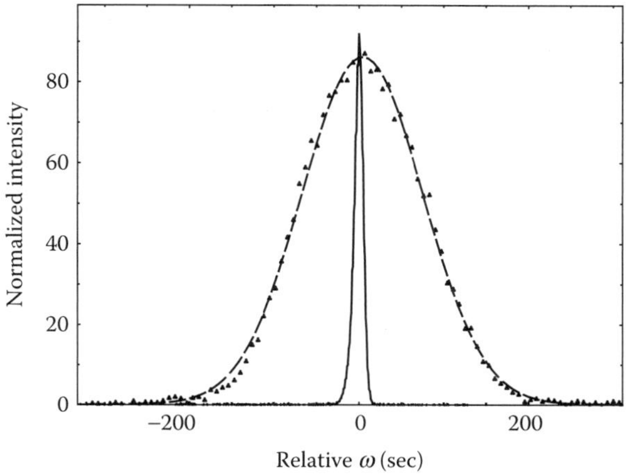

FIGURE 5.2
Triple-axis HRXRD scans of a thick epilayer of $Hg_{1-x}Mn_xTe$ grown by organometallic vapor phase epitaxy on (001) GaAs. Solid line is a $\omega - 2\theta$ scan that measures the strain dispersion in the epilayer; data points are from a rocking curve, measuring the mosaic width; and the dashed line is a Gaussian fit to the rocking curve (mosaic distribution). (From T.A. Lafford et al., *Phys. Stat. Sol.*, 195, 265–270, 2003. With permission.)

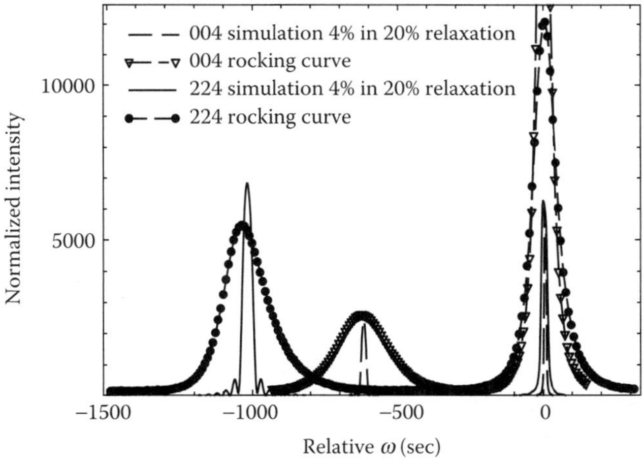

FIGURE 5.3
004 and 224 double-axis rocking curves from a 1-μm-thick layer of $In_{0.04}Ga_{0.96}As$, doped with 5×10^{18} cm^{-3} Si. A 20% relaxation of the layer with respect to the substrate was found to give a good fit to both reflections, as shown in the simulated curves. (From Tanner, B.K. et al., *J. Phys. D: Appl. Phys.*, 36, A198, 2003. With permission from IOP Publishing, Ltd.)

the beam covers a significant area of sample, the Bragg angle position changes across the beam and different parts of the wafer diffract successively as the sample is rotated. The evidence for this interpretation comes from the broadening of the substrate peak to the right, which was sharp before the epilayer was grown. However, the layer peak is wider than the substrate peak, and that extra width is associated with the mosaic introduced by the

Skew-symmetric diffraction

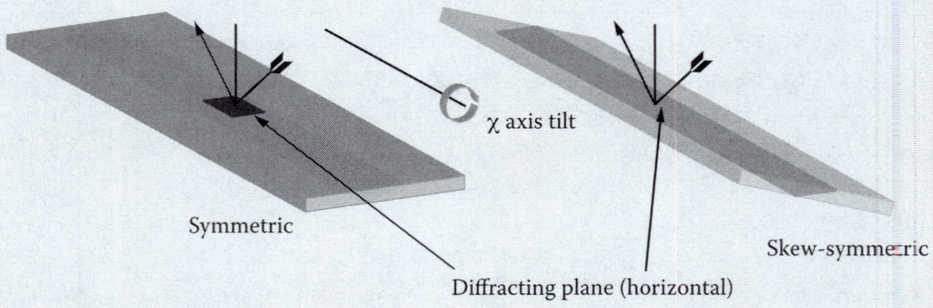

In skew symmetric x-ray diffraction, the specimen is tilted about the X axis until an oblique plane comes into the diffracting position. The angle of the incident and diffracted beams to the surface is thus reduced. This permits the measurement mosaic spread in the plane of the surface (twist mosaic) and also a component of lattice parameter parallel to the surface.

misfit dislocations created as the layer partially relaxed. Subtraction of the two in quadrature enables the mosaic to be measured, provided that we assume no strain distribution is present. If this is not the case, we must resort to the triple-axis configuration to separate tilts and dilations.

The dislocation density and configuration in GaN-based epitaxial systems have a significant effect on the optical emission. These materials are so highly mismatched to their substrate materials that the misfit dislocation density is extremely high in the interface region. However, transmission electron microscopy and high-resolution x-ray diffraction studies have demonstrated clearly that it is the threading edge dislocations that determine the in-plane mosaic and threading screw dislocations the out-of-plane mosaic. Almost all of these threading dislocations run parallel to the film normal.

Measurement of the out-of-plane, so-called *tilt*, mosaic of (00.1) oriented GaN films is very straightforward and is obtained directly from the broadening to the width of the Bragg reflection 00.n rocking curve in a standard surface symmetric high-resolution setting. Determination of the in-plane, so-called *twist*, mosaic requires the use of hk.l diffraction planes that are inclined to the surface. The geometry of the skew symmetric reflections is shown in the inset box. In a skew symmetric reflection, both tilt and twist mosaics contribute to the rocking curve width, and thus a single skew symmetric reflection is not sufficient to separate the two components. It is necessary to measure a series of skew reflections and extrapolate to an inclination angle (ϕ) of 90° (Figure 5.4). The extrapolation is not straightforward and is usually performed, as in Figure 5.4, using a phenomenological model introduced by Srikant et al.[1]

The twist mosaic can be determined from a single measurement if it is possible to measure the rocking curve width of the hk.0 diffraction peak from planes perpendicular to the specimen surface where ϕ = 90°. Due to

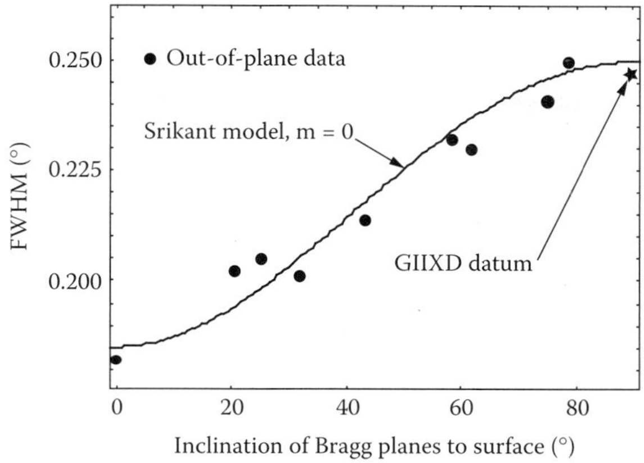

FIGURE 5.4
Full-width-at-half-height maximum of a sequence of skew symmetric reflections from a GaN epilayer on sapphire.

the weak scattering and extensive beam spill-off from the sample, such measurements in the GIIXD geometry have, until recently, been confined to synchrotron radiation sources. However, by using a microfocus x-ray source and focusing optics, very high signals and good signal to noise can be obtained.[2] The resolution is limited by the x-ray optics used and is a trade-off, as always, between intensity and resolution. We commonly use a polycapillary optic with about 0.15° divergence, matched with a Soller slit of similar acceptance, as an analyzer before the detector. However, an asymmetric Ge crystal and Si or Ge analyzer enable an instrumental resolution of 0.01° to be achieved. Alignment is as rapid and as straightforward as for normal high-resolution x-ray diffraction. Figure 5.5 shows the GIIXD rocking

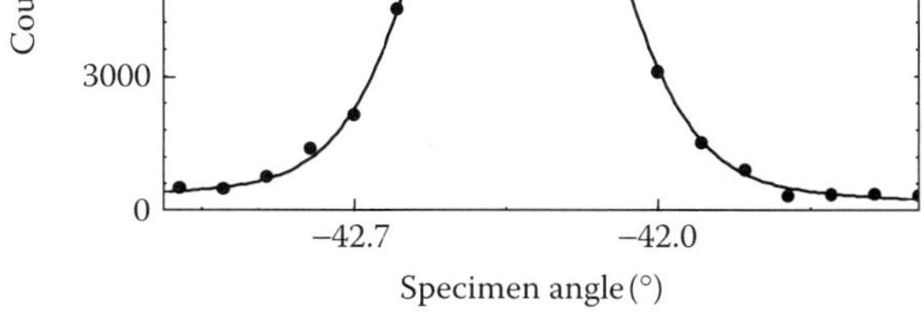

FIGURE 5.5
GIIXD rocking curve of a 0.35-μm-thick GaN epilayer on a sapphire substrate. The solid line is a pseudo-Voigt function fit to the rocking curve. (From T.A. Lafford et al., *Phys. Stat. Sol.*, 195, 265–270, 2003. With permission.)

curve from a 0.35-μm-thick GaN epilayer on sapphire taken on an automated x-ray diffraction tool such as found in many fabs for HRXRD metrology. Total loading, setup, alignment, and measurement time was 5 min. It is worth noting that the GIIXD datum point (filled star in Figure 5.4) is in excellent agreement with the extrapolation from the skew symmetric reflection sequence (filled circles).

5.4 Summary

- XRD peak widths are inversely proportional to crystallite thickness in the direction of the diffraction vector.
- This may be used to measure grain size in thin polycrystalline films from diffraction line broadening.
- Grain morphology may sometimes be deduced.
- Microstrain also broadens peaks, but this broadening increases with scattering angle.
- A Williamson–Hall plot of broadening $\Delta(2\theta)\cos\theta$ against $(\sin\theta)/\lambda$ is normally used to separate size and strain broadening and measure them independently.
- Measurement of the tilt mosaic is done directly from a rocking curve (specimen scan) in the HRXRD triple-axis setting.
- Thick, highly mismatched epilayers relax, and the dislocations form into subgrains that are tilted with respect to each other.
- In such layers, the HRXRD rocking curve in the double-axis setting is dominated by the tilt mosaic.
- The twist mosaic can be measured directly by GIIXD or by extrapolation from a series of skew symmetric HRXRD measurements.

References

1. V. Srikant, J.S. Speck, and D.R. Clarke, *J. Appl. Phys.*, 82 (1997) 4286.
2. T.A. Lafford, P.J. Parbrook, and B.K. Tanner, *Appl. Phys. Lett.*, 83 (2003) 5434.

6

Interface Roughness Metrology

A number of electronic and magnetic properties of a thin film depend upon the roughness of its surface and its interfaces with other films. At the levels of roughness involved in semiconductor and magnetic thin films, x-ray scattering is the only nondestructive technique for measurement of the interface roughness of buried interfaces. For surfaces, scanning probe microscopy (SPM) methods work well for measurement of the topological roughness, provided that the length scale of the roughness is not large. In the case where the in-plane correlation length of the roughness is greater than a few tens of nanometers, the SPM data may be severely truncated if long scan ranges are not employed. Specular x-ray reflectivity is the most commonly used method, but there is no information on interface structure parallel to the interface, and the full interface width is measured. To obtain data more directly comparable with, for example, atomic force microscopy, it is necessary to distinguish between the roughness and the grading of an interface, and this can be done by measurement of the diffuse scatter at low angles. Interface roughness can also be deduced from the shape of the tails of the diffraction peak in a Bragg reflection. This is rarely used but can be helpful in certain cases.

6.1 Interface Width and Roughness

Roughening a surface or interface at the scale of the x-ray wavelength reduces its x-ray reflectivity (XRR). Optical reflectivity may remain unaffected by such short-length scale topography, but the effects are profound upon specular XRR profiles. They are influenced by surface roughness as follows:

- Roughness at the top surface affects the decay rate of the reflectivity. For a perfectly sharp interface, the decay is proportional to θ^{-4} (at angles above the critical angle θ_0 for total external reflection), where θ is the incident angle. Roughness in the top surface makes the

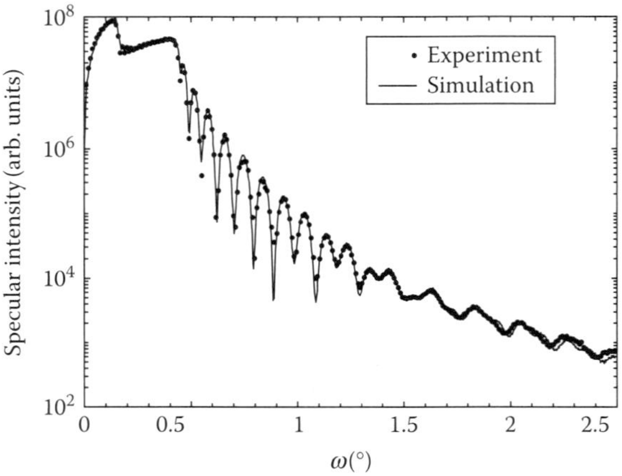

FIGURE 6.1
Specular reflectivity profile of a Co-Cr-Co trilayer at x-ray wavelength of 0.17075 nm. (The structure in Table 6.1 also fits excellently data taken at another x-ray wavelength.) (From T.P.A. Hase et al., *J. Phys. D: Appl. Phys.*, 36, A231, 2003. With permission from IOP Publishing, Ltd.)

profile decay much more rapidly, and measurement and fitting of this curve allow the determination of roughness to about 0.01 nm. The roughness of buried interfaces does not affect the decay profile.

- If a single thin film is present, intensity oscillations (fringes) are seen by interference between waves reflected from the surface and the interface with the substrate. Fringe contrast is a maximum when both these interfaces have equal reflectivity. Roughness of either surface will decrease the fringe contrast. Sensitivity to the roughness of buried layers is typically 0.05 nm.
- If multiple layers are present, fringe contrast will be controlled by the relative reflectivity of each pair of interfaces that cause a component of the interference system. Though this can get quite complex, it can usually be disentangled by modeling, simulation, and fitting methods.

An example of a reasonably complex measurement is shown in Figure 6.1. As in the usual modeling process, the thickness, density, and roughness of each layer were simultaneously optimized to find the best fit between the model simulation and the experimental data.

It is noteworthy that in this case, of a Co-Cr-Co trilayer grown by ultra-high-vacuum evaporation on a Si wafer, inclusion of a 0.95-nm-thick CoO layer on the top surface was essential to get a satisfactory fit to the data. Further, a 0.027-nm layer of native oxide was found to be necessary on the Si substrate surface, as the Si wafer was not baked prior to growth. The parameters deduced from the XRR (Table 6.1) are in very good agreement with those determined from cross-sectional high-resolution transmission electron microscopy (HRTEM) (Figure 6.2).

TABLE 6.1

Comparison of Layer Parameters Deduced from HRTEM and XRR from the Data in Figure 6.1 and Figure 6.2

Sample 1	SiO_2	1st Co Layer	Cr Layer	2nd Co Layer	CoO
HREM thickness (nm)	0.6–1.3	18.9–20.3	0.7–1.4	17.8–20.8	—
X-ray thickness (±0.05 nm)	—	20.8	0.9	21.4	1.65
X-ray interface width (±0.05 nm)	0.27	0.75	0.75	0.75	0.95

FIGURE 6.2
High-resolution transmission electron micrograph of the Co-Cr-Co layer from which was taken the data of Figure 6.1. (From T.P.A. Hase et al., *J. Phys. D: Appl. Phys.*, 36, A231, 2003. With permission from IOP Publishing, Ltd.)

We note that the thickness obtained from XRR is a little higher than that obtained from HRTEM. This probably arises from the fact that the XRR thickness is from the center of the error function describing the interface, while for the HREM images the thickness determined depends on the image contrast. We may thus expect a systematic difference of the order of the interface width.

However, everything said so far about rough interfaces can also be applied to interfaces that are graded in composition or density. As introduced in Section 1.3 and Figure 1.5, both roughness and grading smooth the refractive index (electron density) profile normal to the interface. This is the only material variable that enters into the calculation of the reflectivity of an interface. This has several consequences:

- It is better to think of XRR as determining *interface width*, rather than identifying this purely with roughness.
- If a direct comparison is made between AFM and XRR on samples that have no surface grading of composition or density, agreement

is excellent. However, on the many samples that possess some grading, the XRR-measured interface width will be systematically larger than the AFM-determined roughness. Such samples include polished semiconductor substrates.

However, the effects add in quadrature, and if one is three times the other, the effect of the minor component is only 10%. Thus, beyond this, the minor component has a negligible effect on the total interface width.

- Interdiffusion of material across a sharp interface will decrease its reflectivity (and thus can be measured).

As an example, thin oxide layers such as 1- to 2-nm high-*k* dielectrics are very often found to be graded at the surface. We cannot tell from the specular XRR alone whether this is in fact grading or topographic roughness, so it is reported as an interface width. There may be clues from the material structure. For example, a film with a high interface width may be succeeded by one with a much narrower interface. Since for most materials and growth methods (there are exceptions) interface roughness increases with succeeding depositions, this is a strong indication that grading is responsible for the atypically broad interface of a buried layer.

Roughness is very accurately modeled by x-ray scattering theory, but there are limitations. First, the theory used (Chapters 8 and 9) contains an approximation that breaks down when the roughness is above about 3 to 4 nm. Roughness greater than this can be measured experimentally by XRR, but cannot at present be reliably interpreted. Second, the material structure itself may limit the sensitivity. An interface can be so rough that fringes produced by its interaction with other interfaces are indistinguishable from noise. This is more likely the smaller electron density difference across the interface. This places an upper limit on the roughness that can be measured in a particular specimen. Allowing the model to include roughness greater than this limit may adversely affect its convergence. Fortunately, there are methods to test for this problem in a metrological situation.

Diffuse scatter may, if necessary, be used to distinguish between roughness and grading within an interface width parameter, as shown in the next section.

6.2 Distinction of Roughness and Grading

6.2.1 Measurement by Grazing Incidence Rocking Curves

In Chapter 1, we show two structures in Figure 1.5 that have a similar specular reflectivity, one rough, the other graded. In specular reflectivity, the momentum transfer to the photon on scattering is by definition normal to the macroscopic surface. However, if we consider momentum transfer parallel to the surface, it is obvious that the two structures will scatter differently. The purely rough surface has many regions that are oblique to the macroscopic surface and can scatter in directions that have a component parallel to the surface. This results in off-specular or *diffuse scatter*. A flat, purely graded surface has no such regions and cannot give rise to such scatter. Further study of this question (Chapter 9) results in the following broad conclusions:

- Both roughness and grading cause a decrease in specular reflectivity.
- Roughness causes diffuse scatter. As well as the root mean squared (rms) roughness amplitude, the surface height–height correlation length and the fractal Hurst parameter (jaggedness) are needed to model accurately the shape of the diffuse scatter, but the rms roughness is dominant in determining the total diffuse scatter.
- Grading cannot cause diffuse scatter.

Diffuse scatter is thus a valuable tool to settle problems such as those raised in the last section. A good test for the presence of grading is to take a specular scan and two or three diffuse specimen scans at different fixed detector angles. The roughness deduced from the specular curve should also account for magnitude and distribution of the diffuse scatter if there is no grading.

The simplest method to determine the average roughness of the interfaces of a thin-film material is to record a rocking curve (specimen scan) at fixed detector angle and determine the integrated intensity under the sharp specular peak, I_{spec}, and the total integrated diffuse scatter, I_{diff}. To a first approximation these are related by

$$\frac{I_{diff}}{I_{spec}} = (\exp[Q_z^2 \sigma_t^2]) - 1 \tag{6.1}$$

where

$$Q_z = \left(\frac{2\pi}{\lambda}\right) \sin \Phi \tag{6.2}$$

Φ being the detector angle with respect to the incidence beam. From this single measurement, the rms roughness amplitude σ_t can be measured independently of the interface grading. A combination of the XRR profile and a rocking curve often is sufficient to separate the compositional grading and topological roughness.

The above method is an approximation that becomes less accurate as the in-plane length scale of the roughness decreases. It is advisable, if possible, to take rocking curves at different detector angle settings to check for convergence of the roughness value obtained. For very rough surfaces, this may not be practical as the true specular peak rapidly disappears as the detector angle is increased. Although excellent scans of the diffuse scatter may be obtained for high detector angles, there will be no specular peak with which to compare the diffuse scatter.

6.2.2 Measurement by Off-Specular Specimen Detector Scans

A second key method, particularly suited to determination of the type of roughness present, is to perform the equivalent of an XRR scan, but with the sample angle initially offset a small amount from the zero position. The sample and detector are incremented in the ratio of 1:2. This measurement is crucial for accurate determination of layer parameters from the specular scatter, as the diffuse scatter in the forward direction, approximately measured by the offset scan, must be subtracted from the measured XRR curve to obtain the true specular scatter.

From a thin-film structure, interference fringes are often seen in the off-specular coupled scan. This represents coherent diffuse scatter and gives a measure of the degree of conformality between the roughness on different interfaces. To obtain a quantitative measurement of the fraction of roughness that is conformal, it is necessary to fit the data to a model structure. Similarly, when there is conformality between interfaces in a period multilayers structure, low-angle Bragg peaks appears in the offset scans.

6.3 Roughness Determination in Semiconductors

A good example of a self-organized interface structure, in which we can apply the method of Section 6.2.2, appears in SiGe epitaxial superlattice growth.[1] A transmission electron microscopy (TEM) image is shown in Figure 6.3. The interesting interface morphology results arise because compressively strained epilayers, in this case the SiGe on Si layers, have a tendency to relax via stress-driven surface diffusion, which results in the formation of surface undulations. The morphology depends on a variety of factors, including the composition and thickness of the layers and the growth temperature. Tensile epilayers, in the Si on SiGe layers, have a reduced tendency to form surface undulations. This is shown nicely in the TEM image, which shows undulated SiGe layers and comparatively planar Si layers. Specular XRR (confirmed later by diffuse scatter TEM) showed that the SiGe layers have a higher rms roughness than the Si layers.

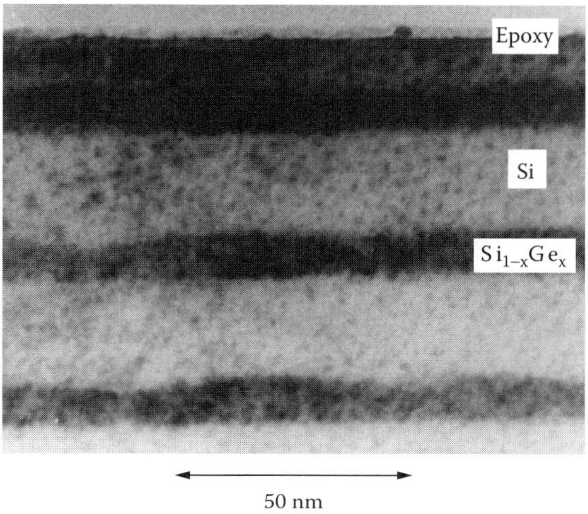

FIGURE 6.3
Cross-sectional TEM image of the Si-SiGe superlattice showing the roughness of the two interfaces. The undulations on the SiGe layers match even though the Si layers have a much flatter interface. (From Powell, A.R. et al., *Semiconductor Sci. Technol.*, 7, 627, 1992. With permission from IOP Publishing, Ltd.)

What is interesting in the TEM image is that the positions of the peaks and valleys of the undulating SiGe layers appear to correlate to positions in the underlying SiGe layers even though the Si layers are not undulated. This is due to the propagation of local strain fields from the bottom to the top of the super lattice structure. This vertical correlation should therefore be apparent in the diffuse scattering. A longitudinal diffuse scan (away from the specular ridge) together with its best-fit simulation is shown in Figure 6.4. It contains a structure similar to that in the specular scan, thus indicating the presence of vertically correlated roughness. Further work[2] showed that the best-fit horizontal correlation length is comparable to the distance between adjacent peaks and valleys as estimated from the TEM image. The best-fit correlation lengths (horizontal and vertical) and fractal parameter, as always, represent average quantities for the super lattice as a whole.

Within the semiconductor industry, measurement of diffuse scatter is often of more use in process development than in in-line metrology, since (with the exception of diffuse scatter from pores, treated in the next chapter) the low intensities mean that it is a fairly slow method. The computation for analysis also takes significant time with current desktop scale computers. However, once a structure has been analyzed carefully, the bounds of its model parameters can be set with confidence. Specular XRR can then normally be used to determine the parameters and for quality control.

A combination of XRR and diffuse scatter provides a nondestructive alternative to AFM for measuring substrate roughness. However, the major advantage is to use the same combination to distinguish chemical intermixing from topological roughness in buried interfaces.

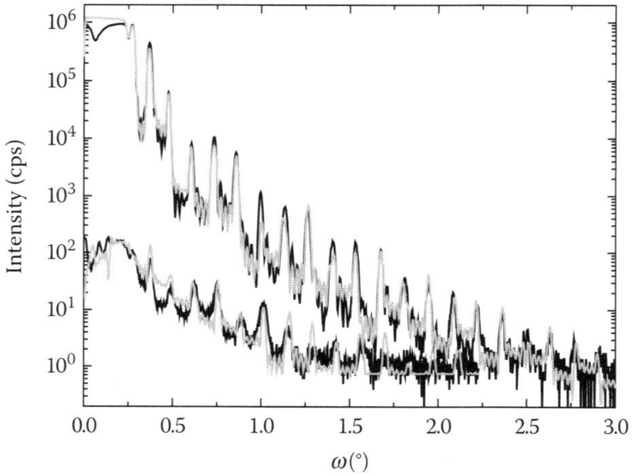

FIGURE 6.4
XRR measurements on the Si-SiGe superlattice. The upper curves show the measured (black) and best-fit simulation (gray) specular scans, and the lower curves the measured (black) and best-fit simulation (gray) longitudinal diffuse scans.

Interfaces in semiconductors are often atomically abrupt, and in this case, the interface width measured in the specular XRR is found to be equal to the amplitude of the topological roughness needed to fit the diffuse scatter. For semiconductor materials that exhibit layer-by-layer growth, the fractal model is not always the best one to fit the diffuse scatter. However, with laboratory sources, the intensity of diffuse scatter data is presently barely sufficient to distinguish between the levels of fit of the various models.

While many Si- and GaAs-based epitaxial semiconductor thin films are of very low roughness, nitride systems have very different characteristics. The roughness level of GaN-based epitaxial layers is usually such that interpretation of the roughness from the specular XRR is not possible. The XRR falls so rapidly with detector angle in such very rough layers that no specular peak is visible in the rocking curve scans, making absolute calibration of the roughness amplitude impossible.

6.4 Roughness Determination in Metallic Films

A wide range of nanometer-scale thickness metallic films are used in the magnetic recording industry. While many of these systems are immiscible, many have some degree of chemical intermixing, and use of diffuse scatter is essential to separate topological roughness from the intermixing. Examples where significant intermixing occurs are Al-Co and Co-Cr; immiscible systems include Fe-Au and Co-Cu. The HRTEM image in Figure 6.2 indicates that the Co-Cr-Co interfaces are diffuse, and thus the interface width

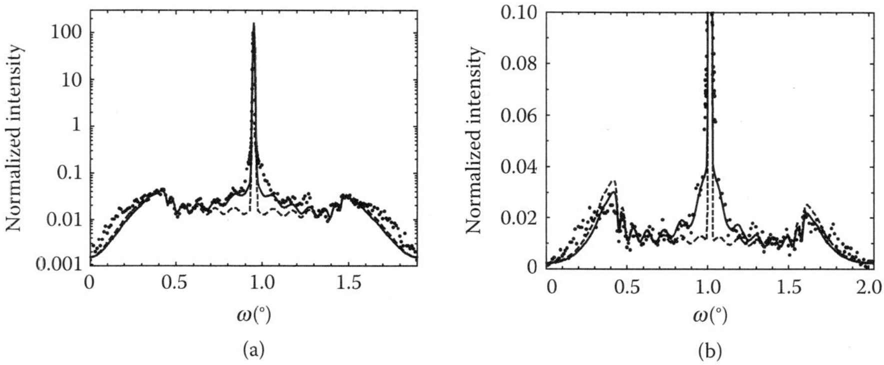

FIGURE 6.5
Diffuse scatter recorded in rocking curves from a Co-Cr-Co trilayer similar to that shown in Figure 6.2. The rocking curves were taken through (a) a Kiessig maximum and (b) a Kiessig minimum in the XRR profile. X-ray wavelength, 0.17075 nm. Solid curves are fits with $\xi = 15.0$ nm and $h = 0.15$. Dashed curves are poorer fits, corresponding to $\xi = 11.0$ nm and $h = 0.25$. (From T.P.A. Hase et al., *J. Phys. D: Appl. Phys.*, 36, A231, 2003. With permission from IOP Publishing, Ltd.)

TABLE 6.2
Fitted Correlated Roughness σ_c, Uncorrelated Roughness σ_u, and Interdiffusion Width Σ for Sample from Which the Data in Figure 6.5 Were Taken

Layer	Thickness (nm)	σ_u (nm)	σ_c (nm)	Σ (nm)
Si	∞	0.05	0.02	0.24
Co	17.85	0.07	0.02	0.64
Cr	1.25	0.22	0.04	0.61
Co	17.25	0.1	0.44	0.66
CoO	1.65	0.4	0.23	0.83

measured by XRR does not correspond to the topological roughness. This is confirmed by the low level of diffuse scatter observed in the rocking curves; that from a similar Co-Cr-Co trilayer is shown in Figure 6.5.

The diffuse scatter from rocking curves taken at different detector angles was fitted to a fractal model with layer thickness and total interface width deduced from the specular scatter, leaving the roughness and intermixing as free parameters, together with the in-plane correlation length, ξ, and fractal parameter, h. (Co-minimization was also done in the above case for a second wavelength.) From the best fit to all the data, the layer parameters shown in Table 6.2 were deduced. The intermixing width Σ is seen to be greater at all interfaces than either the conformal roughness, σ_c, or random roughness, σ_u.

The correlation length, ξ, measured from the x-ray diffuse scatter corresponds quite well to the lateral grain size observed in the HREM images. It is most obvious in the 15- to 20-nm surface undulations visible in the HREM images. While the best fit to the data is a model with very low fractal

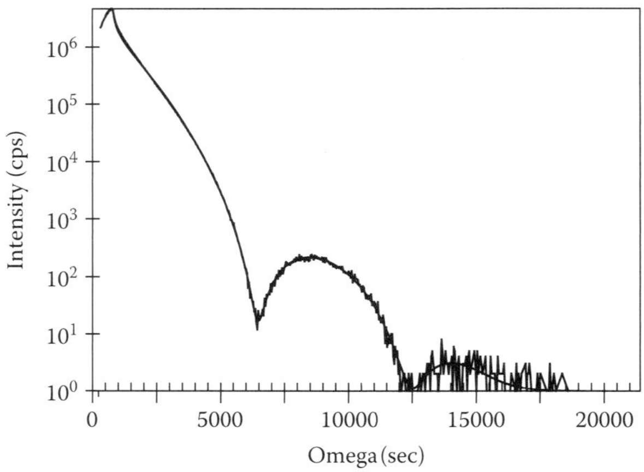

FIGURE 6.6
XRR measurement of a nominally 2.5-nm HfSiON dielectric layer on Si. Both experimental data and the fitted curve are shown as solid lines. Density, thickness, and roughness are obtained from these data. A lower-density layer needed to be included at the top surface in order to obtain a good fit, as is common in such materials. (Courtesy M.A. Quevedo-Lopez, Jeff J. Peterson, P.D. Kirsch, and SEMATECH.)

parameter h, and hence high fractal dimension, there is some doubt as to whether the fractal interface represents a physically realistic model for this system.

The limited intensity available from laboratory sources means that the diffuse scatter from high-quality nanoscale films is very low and difficult to detect. With synchrotron radiation this is not the case, and a number of studies of the propagation of roughness through multilayer films have been published. With a two-dimensional area detector, the diffuse scatter out of the incidence plane can be measured, the so-called grazing incidence small-angle x-ray scattering (GISAXS) technique.

6.5 Roughness Determination in Dielectrics

The roughness as well as the thickness of a high-k dielectric can affect its properties and those of the gate layer. An XRR curve for a 2.5-nm HfSiON layer is shown in Figure 6.6, and the parameters extracted from measurement of two such layers, before and after an annealing process, are given in Table 6.3. This indicates the precision obtained on such measurements. It is not difficult to measure such high-density materials down to ~1-nm thickness with good repeatability, even in a narrow scribe line. (In addition, we may determine the fraction of crystallinity of such layers with the same x-ray tool, though not at high throughput.) It is commonly necessary, as it was in this case, to include a 1- to 1.5-nm layer of low-density material

TABLE 6.3

Determination of Parameters for a Set of HfSiON High-*k* Dielectric Layers

Sample	W18 Anneal 1	W18 Anneal 9	W22 Anneal 1	W22 Anneal 9
Roughness (nm)	0.416 ± 0.007	0.349 ± 0.009	0.448 ± 0.008	0.383 ± 0.005
Density (%)	68.07 ± 0.66	66.20 ± 0.83	70.30 ± 0.99	77.68 ± 1.16
Thickness (nm)	2.502 ± 0.004	2.362 ± 0.004	2.582 ± 0.006	2.590 ± 0.005

Note: R = Density ρ is given as a percentage of bulk density; ρ_{HfSiON} = 13.0 g cm^{-3} (arbitrary, ρ_{HfSiO4} = 13.31 g cm^{-3}).

Courtesy M.A. Quevedo-Lopez, Jeff J. Peterson, P.D. Kirsch, and SEMATECH.

(approximately 1 to 1.5 g cm^3) on the top surface to obtain a good fit. It is not clear whether this is a structural feature of the dielectric or a layer of atmospheric contamination such as an adsorbed water layer.

6.6 Summary

- Specular XRR will determine interface width to approximately 0.01 nm at the surface and 0.05 nm for buried layers. This is fast enough and repeatable enough to be used in-line in a fab.
- Interface width includes both topographic roughness and composition or density grading.
- Grading and roughness can be distinguished by diffuse scatter. This is more suitable for process development than for in-line metrology.
- The nature of the roughness, whether predominantly conformal or random between successive interfaces, can be deduced from different types of scan through the diffuse scatter.

References

1. A.R. Powell, D.K. Bowen, M. Wormington, R.A. Kubiak, E.H.C. Parker, J. Hudson, and P.D. Augustus, *Semiconductor Sci. Technol.*, 7 (1992) 627.
2. M. Wormington, private communication, 2000.

7

Porosity Metrology

Manufacturable films of low-*k* dielectrics are firmly on the International Technology Roadmap for Semiconductors to achieve the performance predictions. In order to reduce the *RC* interconnect delays and cross-talk noise associated with the sub-130-nm technology nodes, copper interconnects must be combined with low-*k* interlayer dielectrics (ILDs) having dielectric constants $k \leq 2$. In order to obtain sufficiently low dielectric constants, pores can be introduced into ILD materials, thereby lowering the average density of the ILD. The parameters that can be measured by x-ray metrology (XRM) are overall porosity, P (defined as the volume fraction of pores), and the pore size, D, and its distribution. The porosity controls the actual dielectric constant, whereas the pore size controls materials properties such as the mechanical strength and integrity, and the action of pore sealant layers. Thickness and roughness of the films are also of importance.

Specular x-ray reflectivity (XRR) provides information on the thickness, roughness, and porosity, while diffuse (nonspecular) XRR yields valuable information about the average pore size and the pore size distribution. Measurement principles for both these methods are outlined in Section 1.2.

7.1 Determination of Porosity

The critical angle for total external reflection depends directly on the electron density at the surface. This translates simply into the measurement of physical density if the chemical composition is known. The critical angle scales as the square root of the material density. Thus, for a porous material, the measured critical angle θ_c will be smaller than that, $\theta_c(0)$, of the matrix material. Assuming that all other material parameters remain constant, the density of the porous material ρ is determined from the density of the bulk material ρ_0 by

$$\rho = \rho_0 \left(\frac{\theta_c}{\theta_c(0)} \right)^2 \tag{7.1}$$

In practice, it is much better to model the whole curve near the critical angle, since the latter is difficult to define repeatably. It is only perfectly sharp for a nonabsorbing material. However, modeling will give an accurate and repeatable value of the density.

If the density ρ_0 of the matrix ILD is known, the porosity P follows, simply defined as

$$P(\%) = (1 - \rho / \rho_0) \times 100 \tag{7.2}$$

An example is shown in Figure 7.1. Two critical angles are observed, one for the substrate and one at a lower angle for the low-density film. In this example, thickness fringes are also seen. The material is a MesoELK™ thin film deposited on silicon. From the best-fit simulation, an average mass density of 0.89 ± 0.02 g/cm^3 was obtained. In the case of spin-on materials, the mass density of the matrix can usually be determined by measuring the critical angle of a thin film of the material deposited without pores, i.e., by adding no porogen during the deposition process. In this case the porosity of the thin film was found to be 53%. The precision of density measurement is normally better than 1% (1σ). In the case of organosilicate materials, which often cannot be produced without some porosity, the density of the matrix must be determined by a complementary technique. The thickness of the thin-film measurement is almost independent of materials properties, and t = 531.1 ± 0.1 nm in this example.

Low-k dielectrics may also be graded in density with depth. Within limits, this may also be handled by XRR and simulation methods. The layer is (computationally) split into lamellae of different densities, and a finite-element modeling approach allows a good approximation to the density profile to be obtained as a staircase function.

In many cases, it is more precise to determine the critical angle from the position of the Yoneda wings in the diffuse scatter. This is relatively sharp, even for materials with high roughness.

7.2 Determination of Pore Size and Distribution

It has been known for many decades that scatter near the incident beam will result from a distribution of small particles. Since a particle is just defined as a region of different electron density, this can as well be a pore as a solid particle. The method of small-angle x-ray scattering (SAXS) is

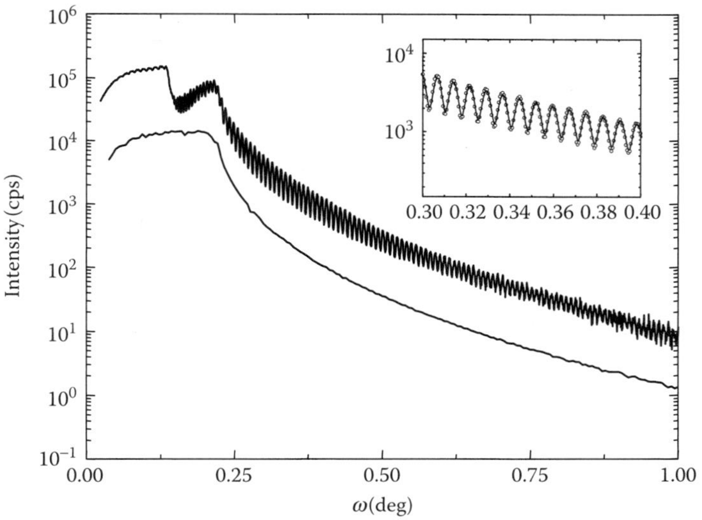

FIGURE 7.1
Specular x-ray reflectivity data from a porous, low-k dielectric thin-film (upper line) sample. The data from a Si substrate is shown for comparison (lower line). The inset shows the agreement between experiment and best-fit simulation (red line) on an expanded scale. (Sample courtesy of John Higgins of Air Products and Chemicals, Inc.)

well developed and has been applied to nanoparticles, clusters, biological fiber bundles, and many other problems. It has also been applied to pores in low-k ILDs. However, there are practical problems, in that the films are generally too thin for optimum scatter, which occurs at a thickness of about $1/\mu$, where μ is the linear absorption coefficient. Since a short-wavelength radiation must be used to penetrate the silicon substrate, and the absorption of a low-k dielectric is, almost by definition, very low, the film should be many times thicker for optimum scatter. The measurements can be made,[1] as can the similar small-angle neutron scattering (SANS),[2] but timescales are far too long for in-fab use.

These problems are to a large extent circumvented by looking at the diffuse x-ray scatter in reflectivity, the grazing incidence small-angle x-ray scattering (GISAXS) geometry. As shown in the box, the scattering volume is large and good signals are obtained. The main limitation, in comparison to the transmission method, is that the lowest scattering angle is equal to the angle of incidence. Thus, there is a cutoff in the information for large pore sizes, which generally occurs for pores 15 to 20 nm in diameter. In the early work on low-k ILDs, some materials had larger pores than this, but there has been a drive to reduce the pore size as much as possible to improve the integrity of the film during processing. Pores currently used in low-k dielectrics are now normally smaller than 5 nm, so the cutoff in information is not a significant problem. The sensitivity to pore size in this regime is high enough for in-line metrology.

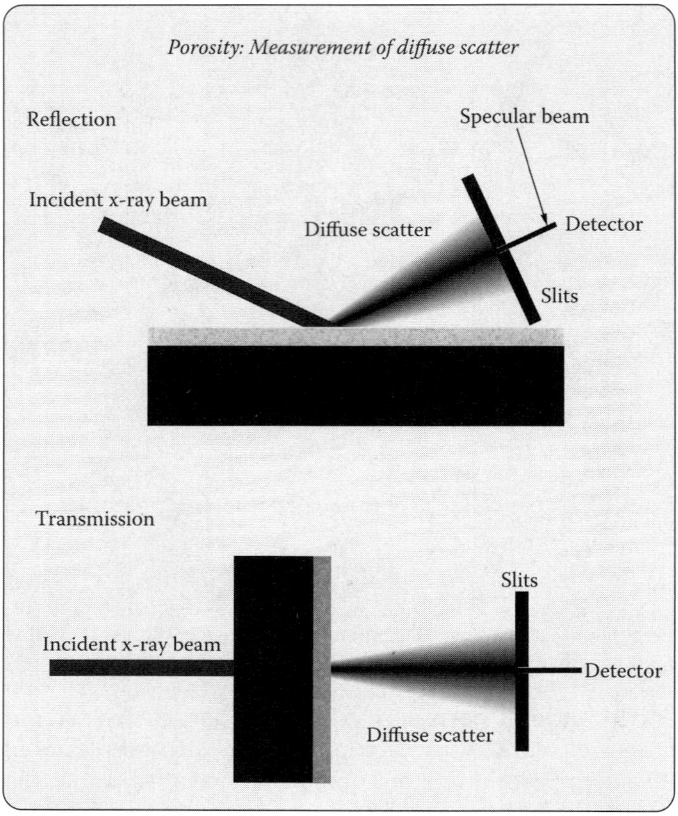

Diffuse XRR may be measured by any method collecting nonspecular x-rays. Our preferred method is to keep the incident angle constant and measure the scatter using a detector (2θ) scan. The advantages of this over, say, an offset $\Omega - 2\theta$ scan are:

- The penetration of the x-ray beam is determined by the incident angle, and is therefore constant throughout the data acquisition scan.
- There is the possibility to study depth variation of pore size distribution by means of scans at different incident angles.
- Rapid, parallel data collection using a position-sensitive detector is easy to achieve.

The theory of diffuse XRR from particles/pores was established by Rauscher et al.[3] The methodology and the models for porosity are taken from those well established for the GISAXS technique. Essentially, the method is to fit the experimental data by simulation from a model, as shown by Omote et al.[4] and also by Wormington and Russell,[5] which this discussion follows. First, one must assume a pore size distribution function. Two models are generally used, illustrated schematically in Figure 7.2:

FIGURE 7.2
Unconnected and connected pore structures. (a) Nonconnected pore structure. Spherical pores with average pore diameter, $\langle D \rangle$. Polydispersity, d, measures the width of pore size distribution. (b) Connected pore structure. Correlation length, ξ, is a measure of the average spacing between regions of phase 1 and phase 2. Average pore dimension is ξ/P, and average wall dimension is $\xi/(1 - P)$.

1. The *polydispersed sphere* model:[6] In this model, the pores are considered to be spherical and surrounded by a matrix material. The pore sizes follow a Gamma distribution that is characterized by two parameters. The first parameter is the mean pore diameter $<D>$ and the second is the polydispersity $d = \sigma^2/\langle D \rangle$, where σ^2 is the variance of the distribution.
2. The *random two-phase* model:[7] In this model one phase consists of the pores and the second phase the surrounding matrix material. This model has an exponential pore size distribution function and is characterized by a single parameter, the correlation length, ξ, which is related to the average pore diameter, $\langle D \rangle$, according to $\langle D \rangle = \xi/(1 - P)$, where P is the volume fraction of pores, i.e., the porosity.

The only way to decide between these models is to see which of them gives the better fit to the data. In this case it was the polydispersed sphere model. Figure 7.3 and Figure 7.4 show the data and the fitting.

To check that there is no significant variation of pore distribution with depth, a series of measurements at varying angle of incidence may be taken. Figure 7.5 shows an iso-intensity map that was produced (on the same sample as for Figure 7.3) from a series of radial diffuse scans measured as the incidence angle, ω, was increased from 0.07 to 0.54° in steps of 0.014°. Variations in the pore size distribution with depth would result in changes in the diffuse intensity distribution with increasing ω. As significant changes are not present in the map, the authors concluded that there is no appreciable variation in the pore size distribution as a function of depth within the

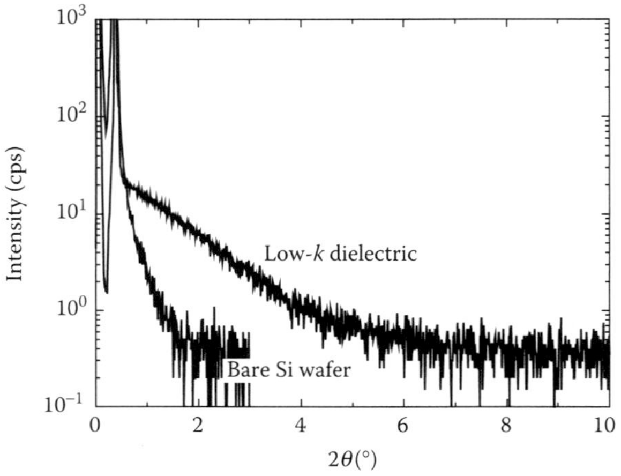

FIGURE 7.3
A detector scan from a low-k dielectric and from a bare silicon wafer, showing the diffuse scatter.

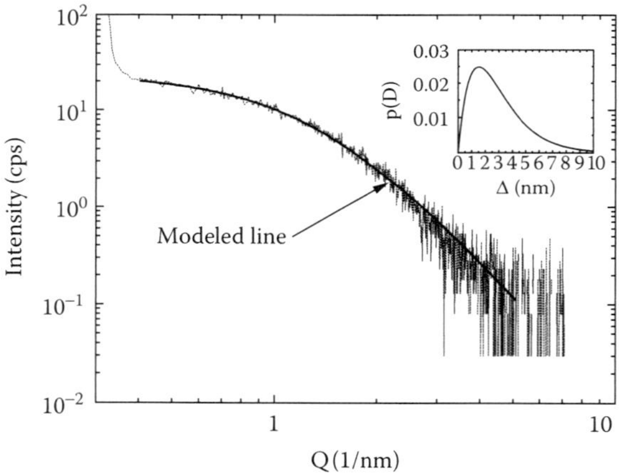

FIGURE 7.4
With the background subtracted, the data are fitted to a theoretical curve (solid line) modeled using a polydispersed sphere distribution (shown in inset). The average pore size was deduced to be $\langle D \rangle = 3.0 \pm 0.3$ nm and the polydispersity factor $d = 0.7$.

porous, low-k sample. The diagonal streak in the map is the specular scatter and the broad, diffuse "hump" centered around $\omega \sim 0.175°$ results from a maximum in the transmitted amplitude when the exit angle of the scattered radiation is equal to the critical angle of the thin film (Yoneda wing, discussed in Chapter 9).

This discussion has gone into some detail in order to bring out an important difference in the modeling of porosity-induced diffuse scatter from, say, the modeling of high-resolution x-ray diffraction (HRXRD) or XRR profiles. In both cases the calculation of the x-ray scatter from the model is highly

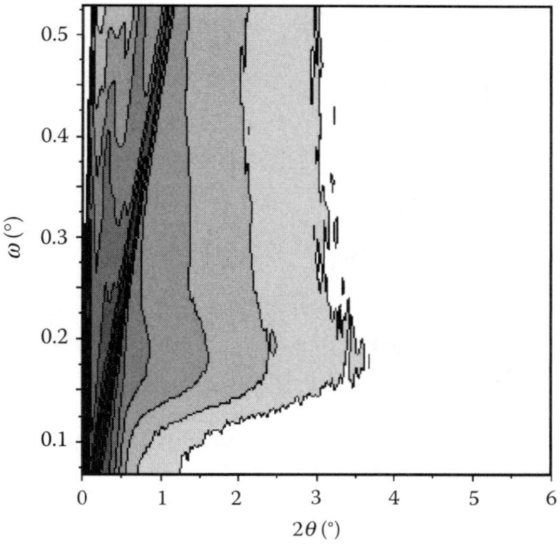

FIGURE 7.5
Iso-intensity map of the diffuse XRR data from a porous low-k dielectric thin-film sample as a function of the incidence angle, ω, and scattering angle, 2θ. Darker corresponds to more intensity scattered from the sample.

accurate. However, in the cases of epitaxial layers and of thin reflecting layers, the model itself is simple and unambiguous. This is not the case for diffuse scatter from pores. The reasons are:

- The diffuse scatter depends not only on the average pore size, but also on its distribution. Several models (polydispersed spheres, random two-phase, exponential) are possible for the distribution, and these give differing numerical results. There may be no *a priori* reason to select one over another, and the selection will then depend on goodness of fit and more advanced analysis using Bayesian statistics. On the other hand, it is likely that particular models will be found appropriate for certain materials and processes.
- Other structural features cause diffuse scatter, in particular surface roughness and (if present) other dispersed particles. Pore scatter usually dominates over roughness scatter except very close to the incident beam, but this is not an infallible rule.
- More complex dispersions, e.g., bimodal or depth dependent, will cause different shapes of the diffuse scatter profile. Since the profile is rather featureless, containing no interference fringes, the sensitivity to such complexity may not be very high; it is not yet established.

Porosity metrology is thus not absolute and traceable in the same sense as thickness metrology. It is dependent upon the model and on the material. For a particular materials system, and a specific process, the variables can be understood by cross-comparison with other techniques such as positron annihilation spectroscopy (PALS) and ellipsometric porosimetry (EP). It may then be used with confidence in process development and quality control.

7.3 Pores in Single Crystals

Most porous materials are nanocrystalline or amorphous, but there is an interesting exception in the case of silicon. Using electrochemical methods, layers of porous material up to typically 30 μm thick can be prepared on the surface of silicon wafers. The porosity is determined by the current density during the anodic reaction and can vary from 30 to 70%. Pore radii are typically 3 to 6 nm.

XRR can be used, as for the low-*k* dielectric material, to determine the porosity, and diffuse scatter measurements enable the pore size to be modeled. Barla et al.[8] used HRXRD to measure the strains associated with the porosity and observed that the porous layer behaved as a well-defined epitaxial single crystal of reduced density with respect to the substrate. As a result, the wafers were significantly bowed. The HRXRD rocking curves showed that the lattice parameter measured normal to the wafer surface, i.e., in the symmetric scattering geometry, was greater than that of the silicon substrate. (A well-resolved and sharp Bragg angle from the porous layer was observed at lower angle than that from the substrate.) Barla et al.[8] found that the lattice expansion normal to the surface increased linearly with both increasing porosity and pore radius. A combination of XRR and HRXRD is thus particularly powerful for obtaining independently the lattice strain and the porosity of porous silicon. It should be noted that the Poisson ratio is affected by porosity and may be unknown.

7.4 Summary

- Measurement of porosity is straightforward from XRR if the matrix density is known.
- If the matrix density is not known, the XRR data will require calibration, but will be reproducible between samples if the matrix density is constant.
- The depth profile of porosity may be measured from the specular reflectivity profile.

- The diffuse scatter measured in grazing incidence reflectivity shows sensitivity to the mean pore size and its distribution.
- Pore size distribution may be measured by fitting a model to the diffuse scatter.
- The popular models are polydispersed sphere and random two-phase. There is usually enough sensitivity in the data to distinguish between these models.
- Features such as surface roughness, bimodal or multimodal distributions, or depth grading of pore size will interfere with the analysis. Detailed verification is required in each of these complex cases to determine whether x-ray diffuse scatter alone provides sufficient information for metrology.

References

1. W.-L. Wu, W.E. Wallace, E.K. Lin, G.W. Lynn, C.J. Glinka, E.T. Ryan, and H.M. Ho, *J. Appl. Phys.*, 87 (2000), 1193–1200.
2. E.K. Lin, H.J. Lee, B.J. Bauer, H. Wang, J.T. Wetzel, and W.L. Wu, *Low Dielectric Constant Materials for IC Applications*, Springer Series in Advanced Microelectronics, P.S. Ho, J. Leu, and W.W. Lee, Eds., Springer Publishing, Berlin, 2003, pp. 75–93.
3. M. Rauscher, T. Salditt, and H. Spohn, *Phys. Rev. B*, 52, (1995) 16855.
4. K. Omote, Y. Ito, and S. Kawamura, *Appl. Phys. Lett.*, 82, (2003) 544.
5. M. Wormington and C. Russell, *ULSI 2003 AIP Conference Proceedings*, Vol. 683, Issue 1, pp. 651–655.
6. W.L. Griffith, R. Triolo, and A.L. Compere, *Phys. Rev. A*, 35 (1987) 2200.
7. P. Debye, R. Anderson, and H. Brumberger, *J. Appl. Phys.*, 28 (1957) 679.
8. K. Barla, R. Herion, G. Bomchil, J.C. Pfister, and A. Freund, *J. Crystal Growth*, 68 (1984) 727.

Part 2

The Science

8

Specular X-ray Reflectivity

8.1 Principles

The refractive index of x-rays in matter is less than unity.* The consequence of this is that for low angles, total external reflection occurs at the air–material interface. Just as for conventional optics, where $n > 1$ and total internal reflection occurs, there is a critical angle at which total reflection ceases and the wave penetrates the material.

Because the refractive index for x-rays is very close to unity, it is conventional to write the (complex) refractive index **n** in the form

$$\mathbf{n} = 1 - \delta - i\beta \tag{8.1}$$

The δ and β terms are determined by dispersion and absorption, respectively, β being directly proportional to the linear absorption coefficient. A straightforward calculation shows that this can be written as

$$\mathbf{n} = 1 - \frac{\lambda^2 r_e}{2\pi} \sum_a (Z_a + f_a' + if_a'')N_a \tag{8.2}$$

where λ is, as usual, the x-ray wavelength and r_e (the classical electron radius) is given by

$$r_e = e^2 / 4\pi\varepsilon_0 mc^2 \tag{8.3}$$

where m is the electron mass, e the electronic charge, c the speed of light, and ε_0 the permittivity of free space. Z_a is the atomic number of species a in the material (i.e., the number of electrons per atom), N_a is the number of

* Although this means that the phase velocity of x-rays in matter is greater than the speed of light in vacuum c, the group velocity, with which information can be transmitted, always remains less than c. Einstein's theory of special relativity is not violated in x-ray metrology.

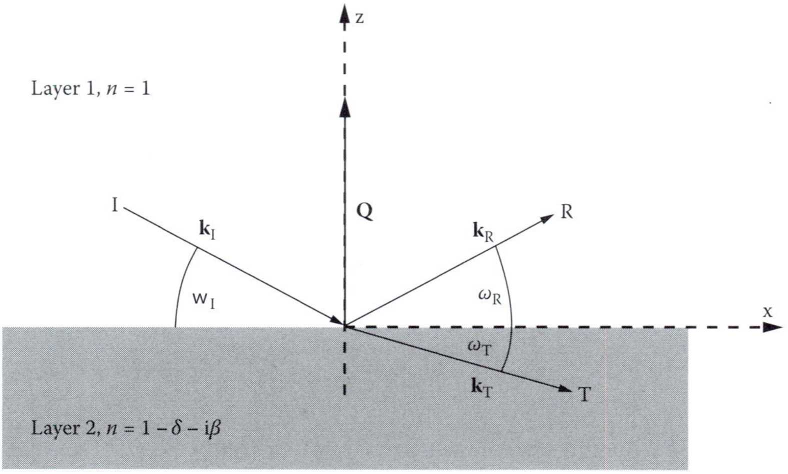

FIGURE 8.1
Wave vector and ray directions associated with an electromagnetic wave incident on an ideal interface between materials of different electron densities.

atoms of species a per unit volume, and f_a' and f_a'' are the real and imaginary parts of the dispersion correction to the scattering. The values are tabulated in sources such as the *International Tables for X-Ray Crystallography*. To the x-ray metrologist, Equation 8.2 is particularly important, as it shows that the refractive index depends primarily on the atomic number and the electron density. It also illustrates why the nucleus is unimportant; the equivalent equation for protons is a factor of approximately 2000 lower through the mass in Equation 8.3. X-ray scattering is thus essentially sensitive to electron density only.

In the x-ray region of the electromagnetic spectrum the values of both δ and β are small and positive, of the order of 10^{-6} and 10^{-8}, respectively, giving a refractive index only very slightly less than unity. As a consequence, x-ray radiation incident from air on to a material interface sees the material as being optically less dense and is refracted away from the surface normal, as illustrated in Figure 8.1.

As the incidence angle with the surface θ_I is decreased, the angle between the transmitted wave and the surface θ_t also decreases. As it is always less than θ_I, there is a critical angle θ_c at which the transmitted wave travels parallel to the surface. For incidence angle below θ_c, there is total external reflection. The value of the critical angle can be derived very straightforwardly from Snell's law.

The complex amplitude E_j of each of the x-ray beams shown in Figure 8.1 can be expressed as a plane wave:

$$E_j = C_j e^{i\mathbf{k}_j \cdot \mathbf{r}} \tag{8.4}$$

where $\mathbf{k}_j$ ($|\mathbf{k}| = 2\pi/\lambda$) is the wave vector. The first term, C_j, where $j = I, R$, and T for the incident, reflected, and transmitted waves, respectively, defines the amplitude and the exponential term defines the phase of the wave. From the fundamental definition of refractive index, we have for all media

$$\frac{|\mathbf{k}_I|}{n_{layer1}} = \frac{|\mathbf{k}_T|}{n_{layer2}} = \frac{|\mathbf{k}_R|}{n_{layer1}} = \frac{|\mathbf{k}|}{n_{vacuum}} \tag{8.5}$$

Exactly as for the case of visible light optics, and explained in many textbooks on classical electrodynamics, the wave must be continuous at the interface; the amplitude and gradient of E_j must match at the interface for each of the plane waves. This gives

$$C_I + C_R = C_T \quad \text{and} \quad C_I\mathbf{k}_I + C_R\mathbf{k}_R = C_T\mathbf{k}_T \tag{8.6}$$

When the surface component of the plane wave at the boundary is considered explicitly in Equations 8.5 and 8.6 the familiar form of Snell's law is obtained:

$$n_{layer1}\cos\theta_i = n_{layer2}\cos\theta_t \tag{8.7}$$

If the wave is incident from vacuum or (to an extremely good approximation) air, $n_{layer1} = 1$, then the critical angle (where $\theta_t = 0$) is defined as

$$\cos\theta_c = n_{layer2} \approx 1 - \frac{\theta_c^2}{2}\ldots\ldots \tag{8.8}$$

If it is assumed that there is no absorption, $\beta = 0$, then a simple rearrangement yields an expression for the critical angle as

$$n_{layer2} = 1 - \delta \tag{8.9}$$

Thus,

$$\theta_c = \sqrt{2\delta} \tag{8.10}$$

Equation 8.10 is immediately useful to the x-ray metrologist as it shows that measurement of the critical angle yields directly the electron density in the near-surface region. Depending on what else is known, this can be used to measure either the chemical density, the porosity, or the chemical composition. It is important to note here that of these, two must be fixed and only one can be the independently determined quantity.

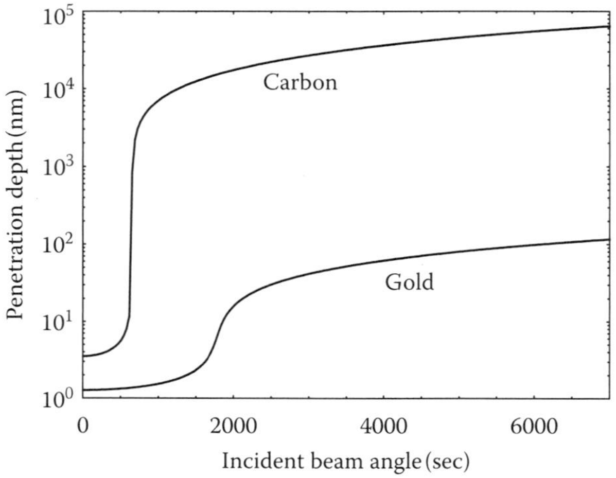

FIGURE 8.2
Penetration depth (the distance at which the intensity falls off by $1/e$ from its value at the surface) as a function of incident beam angle for a carbon (upper line) and gold (lower line) layer, calculated for a wavelength of 0.138 nm. (Courtesy B.D. Fulthorpe, Ph.D. thesis, Durham University.)

For incident beam angles below the critical angle, solution of the above equations results in a value for the wave vector in the medium, $\mathbf{k}_T$, that is imaginary. Physically, this results in an exponential fall-off of the intensity with distance below the surface. The fall-off is fast and there exists a so-called evanescent wave, which penetrates only a small distance into the material, even though the x-rays are totally reflected from the interface. The simulated penetration depth as a function of incident beam angle is shown in Figure 8.2 for carbon and gold. The depth penetration of this evanescent wave is primarily dependent on the electron density of the near-surface material. This is illustrated in Figure 8.3, which shows the amplitude of the electric field as a function of depth and angle for these two materials.

The fluorescence yield as a function of angle can be calculated directly from the electric field amplitude, and hence the fluorescence signal as a function of angle can be quantified. Grazing incidence x-ray fluorescence (GIXRF) spectroscopy has a very high sensitivity to depth in the range just beyond the critical angle, where the x-ray penetration varies very rapidly with angle. The shape of the fluorescence signal as a function of angle can be used to determine the position of impurities or very thin layers with respect to the surface. While total external XRF (TXRF) is used extensively within the semiconductor industry to enhance the signal/noise of fluorescence from impurities at or very near the surface, the depth sensitivity of GIXRF has been exploited very little so far.

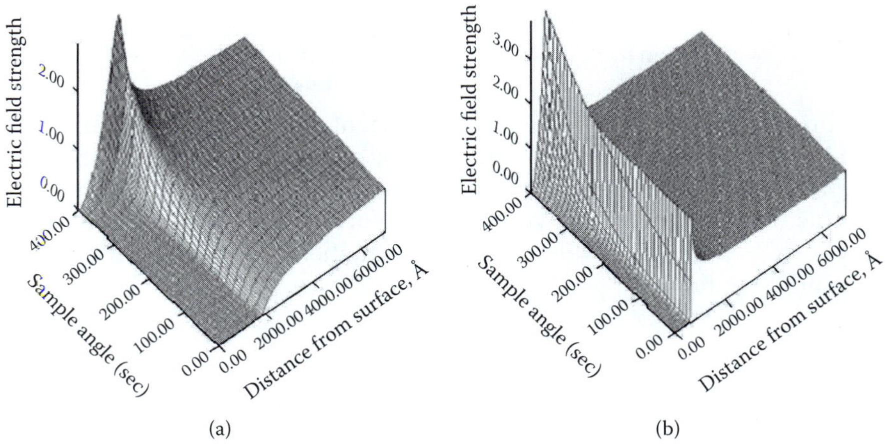

FIGURE 8.3
Electric field as a function of incidence angle and distance from surface, $\lambda = 0.138$ nm.[1] (a) Gold. (b) Carbon. (Courtesy B.D. Fulthorpe, Ph.D. thesis, Durham University.)

8.2 Specular Reflectivity from a Single Ideal Interface

The amplitudes of the transmitted and reflected waves can be determined by the conditions for continuity of the waves at the interface, resulting in the so-called Fresnel equations:

$$F^R_{layer1} = \frac{k_{I,Z} - k_{T,Z}}{k_{I,Z} + k_{T,Z}}, \qquad F^T_{layer1} = \frac{2k_{I,Z}}{k_{I,Z} + k_{T,Z}} \tag{8.11}$$

for the amplitudes of the reflected wave F^R_{layer1} and the transmitted wave F^T_{layer1}.

Here,

$$k_{I,Z} = -k_{R,Z} = |k_{I,Z}|\sin\theta_I = kn_{layer1}\sin\theta_I \tag{8.12}$$

and

$$k_{T,Z} = |k_{T,Z}|\sin\theta_T = kn_{layer2}\sin\theta_T \tag{8.13}$$

The Fresnel amplitudes can therefore be expressed in terms of the incident and transmitted angles only, such that

$$F^R_{layer1} = \frac{\sin\theta_I - \sin\theta_T}{\sin\theta_I + \sin\theta_T} \approx \frac{\theta_I - \theta_T}{\theta_I + \theta_T}, \qquad F^T_{layer1} = \frac{2\sin\theta_I}{\sin\theta_I + \sin\theta_T} \approx \frac{2\theta_I}{\theta_I + \theta_T} \tag{8.14}$$

In principle, there are two sets of coefficients, one for each polarization state. However, because the scattering angles are small, the polarization dependence can be neglected in grazing incidence reflectivity.

Let us consider the air (or vacuum) interface with a single solid surface. Now $n_{layer1} = 1$ and $n_{layer2} = n$. Then from Equations 8.7 and 8.9 in the low-angle limit we have

$$1 - \frac{\theta_I^{\,2}}{2} = n(1 - \frac{\theta_T^{\,2}}{2}) = (1-\delta)(1 - \frac{\theta_T^2}{2}) \tag{8.15}$$

Thus,

$$\theta_T^2 = \theta_I^2 - 2\delta \tag{8.16}$$

$$F^R_{layer1} = \frac{\theta_I - \theta_T}{\theta_I + \theta_T} = \frac{1 - \sqrt{1 - \dfrac{2\delta}{\theta_I^2}}}{1 + \sqrt{1 - \dfrac{2\delta}{\theta_I^2}}} \approx \frac{\delta}{2\theta_I^2} \tag{8.17}$$

The specular scattered intensity I is given by $I = |F^{R*}F^R|$. In the region beyond the critical angle this falls with the incidence angle, θ_I, to the inverse fourth power, i.e.,

$$I \propto \theta_I^{-4} \tag{8.18}$$

This variation in the mirror-reflected intensity is shown in a simulation in Figure 8.4 and compared with experimental data from a polished Si surface. Note that sometimes the product $I\theta_I^4$ is plotted; as for such a perfectly smooth and nongraded interface, this product is constant as the incidence angle is increased. (In practice, this product falls due to the effect of the nonzero width of the interface and then rises as background noise starts to become significant with respect to the specular reflectivity.)

8.3 Specular Reflectivity from a Single Graded or Rough Interface

As seen above, real surfaces are neither ideally flat nor of constant composition on either side of the interface. There is some width to the interface

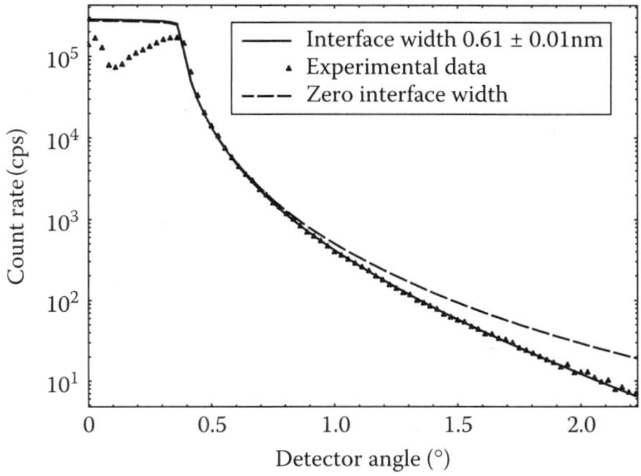

FIGURE 8.4
Experimental and calculated specular reflectivity from a polished Si substrate. Two simulations are shown, one for an ideal, abrupt interface and the other for an interface width at the surface of $\sigma = 0.61$ nm.

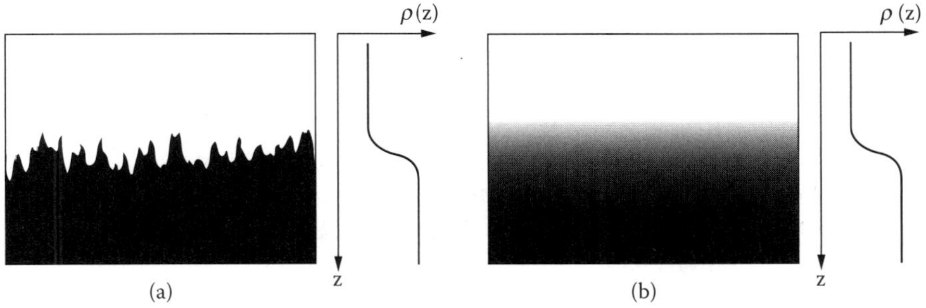

FIGURE 8.5
Electron density, $\rho(z)$, averaged parallel to the interface plane across (a) a topologically rough and (b) a compositionally graded interface.

and this affects the specularly reflected intensity. A rough or compositionally graded interface can be modeled as a series of perfectly flat interfaces with a distribution about some average position. We can choose various distributions depending on the nature of the interface, but the most common practice is to model the interface electron density profile, averaged parallel to the surface, as an error function, as in Figure 8.5. (Note that for specular reflectivity, we cannot distinguish between a topologically rough and a compositionally graded interface. We will come back to this important point later.) The displacement distribution is then a Gaussian, and the standard deviation σ describes the average interface width.

The graded interface profiles can be incorporated into the x-ray scatter when the reflectivity is small, i.e., within the so-called Born approximation,

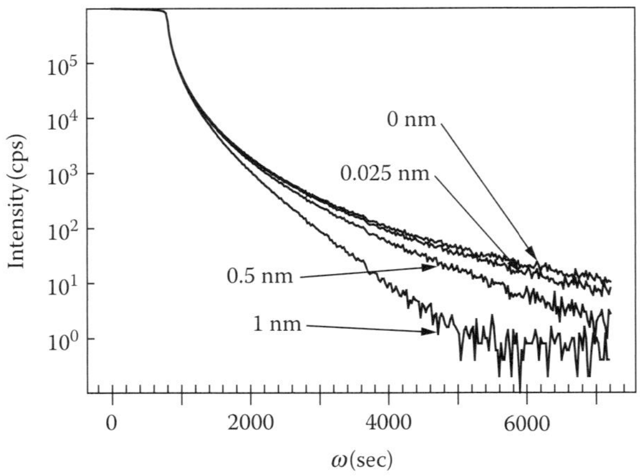

FIGURE 8.6
Simulations of single Si surface of interface width (roughness) varying from 0 to 1 nm.

by modifying the Fresnel reflection amplitude with a phase factor varying over the interface. Explicity, this is

$$F^R_{layer} = F^R_{ideal-layer} \int_{-\infty}^{\infty} \left[\frac{-1}{\rho_{-\infty}} \frac{d\rho(z)}{dz} \right] e^{-iQ_z z} dz \tag{8.19}$$

where $\rho_{-\infty}$ is the electron density deep in the substrate material, $\rho(z)$ is the density normal to the interface, and Q_z is the component of the scattering vector (defined below) normal to the interface. For $\rho(z)$ as an error function, the Fresnel reflectivity is then modified by a so-called static Debye–Waller factor into

$$F^R_{layer} = F^R_{ideal-layer} \exp\left[-\tfrac{1}{2} (Q_z^2 \sigma^2) \right] \tag{8.20}$$

where σ is the root mean squared (rms) interface width. A finite interface width results always in a more rapid fall in intensity than that predicted by Equation 8.18. Examples of the intensity fall for different values of σ are shown in Figure 8.6.

As evident in Figure 8.5, the electron density variation, averaged within the interface plane, is identical for topologically rough and compositionally graded interfaces. Measurement of the specular reflectivity does not distinguish between these two situations. The reason can be seen in Figure 8.1. There, the vector **Q**, known as the scattering vector, is the vector difference between the incident reflected wave vector $\mathbf{k_R}$ and the incident wave vector $\mathbf{k_I}$:

$$\mathbf{Q} = \mathbf{k_R} - \mathbf{k_I} \tag{8.21}$$

In this situation, where the scattered wave is the mirror-reflected wave with $\theta_I = \theta_R$, **Q** is directed along z, the normal to the interface. There can then be no information in the scattered wave about the structure of the interface within the interface plane. Specular reflectivity measures a total interface width and does not distinguish between compositional grading and topological roughness.

The equation works well when the scattering is weak, in the so-called Born approximation, but is not accurate close to the critical angle. Névot and Croce[1] derived a more general form describing the modification to the reflection coefficient between two layers, l and $l-1$, caused by a nonideal interface:

$$F_l^R = F_{l(ideal)}^R \exp\left(\frac{-Q_z^l Q_z^{l-1} \sigma_l^2}{2}\right) \tag{8.22}$$

where Q_z^l and Q_z^{l-1} are the surface-normal components of the scattering vectors in the layered media l and $l-1$, respectively.

An example of the fit to the modified Fresnel reflectivity of the specular reflectivity from a polished Si substrate surface is shown in Figure 8.4. Here, we have modeled the interface width equivalent to an rms roughness of 0.61 ± 0.01 nm. We see that the specular x-ray reflectivity provides a precise metrological tool for measurement of the interface width of polished surfaces. However, as emphasized before, the measurement is of the total interface width, and when compared with an atomic force microscopy measurement, the value of σ is normally significantly higher. Even at polished surfaces of inert materials such as oxide glasses, there is normally a compositional grading.[2] In the case of GaAs, the interface width increases with storage time.[3] For many materials there is a native oxide layer, typically 2 nm thick, that forms on the top surface. This gives rise to low-contrast fringes in the specular reflectivity, typically 1° period (in specimen angle).

Note that beyond a surface roughness of about 3 nm, the theory begins to break down and the values of interface width determined from automatic fitting programs should not be taken as absolute. While the specular reflectivity can still be used qualitatively, it should be with caution and always after subtracting the forward diffuse scatter as discussed in Chapter 9.

8.4 Specular Reflectivity from a Single Thin Film on a Substrate

As illustrated in Figure 8.7, when a film of different electron density to the substrate is present, partial reflection occurs at both the top and bottom

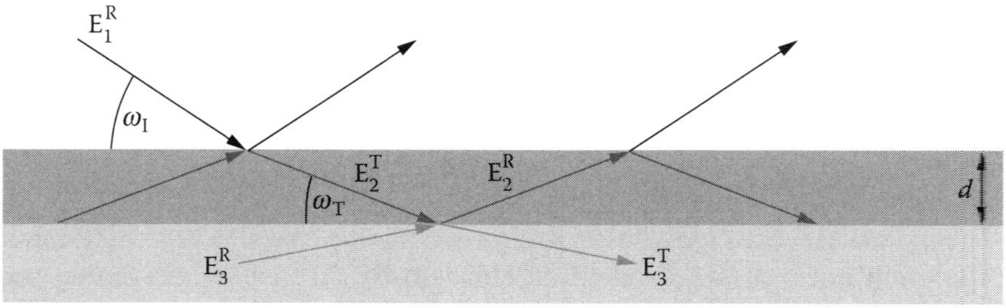

FIGURE 8.7
Wave vectors and ray paths for an electromagnetic wave incident on a thin film of refractive index intermediate between the substrate and air.

interfaces. Interference occurs between the waves, resulting in interference fringes as the optical path difference between them is changed by varying the incidence angle. The fringes are popularly known as Kiessig fringes, after their discoverer. The condition for constructive interference is that the path difference must be an integral number of wavelengths. Thus, we have the same condition as the Bragg law for crystal diffraction, except that now d is the distance between the top and bottom of the layer, rather than the interatomic spacing. From Figure 8.7 we see that this path length is $2d\ \sin\theta_T$, and hence

$$n\lambda = 2d\ \sin\theta_T \tag{8.23}$$

As the incidence angle is changed, the intensity oscillates with a period $\Delta\theta_R$ given by

$$\Delta\theta_T = \lambda/2d \tag{8.24}$$

Through Equations 8.10 and 8.16 we can determine the period of the interference fringes as a function of incidence angle θ_I. Direct inspection of Equation 8.16 shows that for incidence angles typically greater than twice the critical angle, the fringe spacing is approximately

$$\Delta\theta_I \approx \lambda/2d \tag{8.25}$$

Thus, highly accurate thin-film thickness measurements can be made with no information required on the material parameters.

An example of such interference fringes from a nominal 5-nm film of Pt on a Si substrate is shown in Figure 8.8. The best fit shown enables us to determine the layer thickness to be 4.90 ± 0.01 nm, with density 84.4 ± 0.7% that of bulk Pt and interface width 0.23 ± 0.005 and 0.45 ± 0.005 nm at the substrate and top surface, respectively.

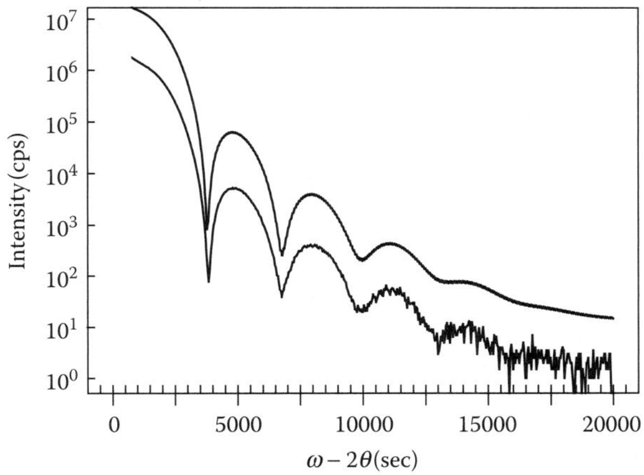

FIGURE 8.8
Specular reflectivity profile of a thin film of Pt on a Si substrate. Best-fit parameters are Pt thickness, 4.94 ± 0.01 nm; rms roughness, 0.24 ± 0.01 nm.

The precise position of fringe number m, derived from Equations 8.10, 8.16, and 8.23, is

$$\theta_m = \sqrt{\left[\frac{m\lambda}{2d}\right]^2 + \theta_c^2} \tag{8.26a}$$

$$\theta_m = \sqrt{\left[\frac{(m+\frac{1}{2})\lambda}{2d}\right]^2 + \theta_c^2} \tag{8.26b}$$

for the cases where the layer electron density is less and greater than that of the substrate, respectively. (There is a phase shift of π on reflection in the latter case that effectively alters the optical path difference by $\lambda/2$.)

For abrupt interfaces, the fringe amplitude depends on the electron density difference between the substrate and the layer. Thus, as illustrated in Figure 8.9, the thickness of SiO_2 on Si is not easy to determine as the fringe amplitude is small compared with, for example, Ta on Si.

However, nonzero interface width also affects the amplitude of the fringes and care should be taken in ascribing changes in fringe visibility to just one or the other of these effects. In the case of a layer of higher electron density than the substrate, for example, Au on Si, increasing width of the substrate–film interface results in reduction in the amplitude of the fringes, though the overall rate of fall of intensity with angle is not affected (Figure 8.10a). Nonzero width of the top surface interface results in both amplitude reduction and a change to the average rate of fall of intensity with angle (Figure 8.10b). The effects of interface width and electron density difference

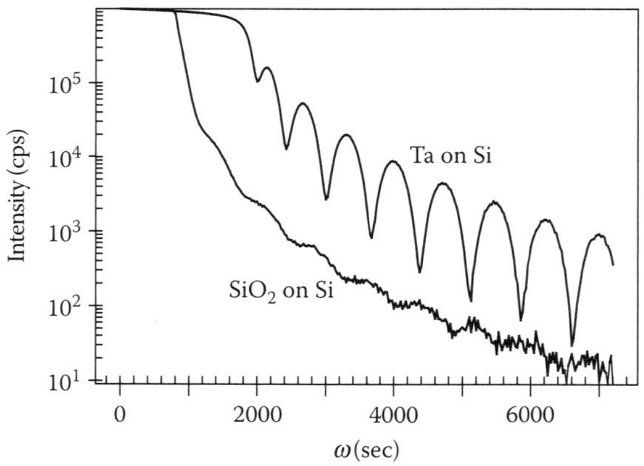

FIGURE 8.9
A (simulated) comparison of SiO_2 on Si and Ta on Si. In each case the layers are 20 nm thick and have zero interface width. $\lambda = 0.154$ nm.

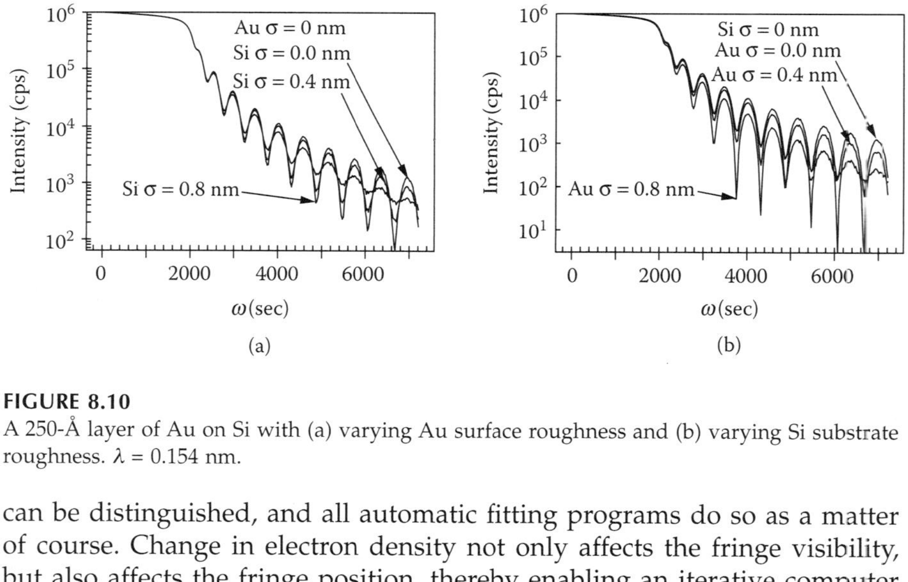

FIGURE 8.10
A 250-Å layer of Au on Si with (a) varying Au surface roughness and (b) varying Si substrate roughness. $\lambda = 0.154$ nm.

can be distinguished, and all automatic fitting programs do so as a matter of course. Change in electron density not only affects the fringe visibility, but also affects the fringe position, thereby enabling an iterative computer program to reach the optimum value for the two parameters.

8.5 Specular Reflectivity from Multiple Layers on a Substrate

The amplitude, R, of the specularly reflected wave from material consisting of many layers of different composition can be calculated using a recursive formalism originally developed by L.G. Parratt for reflection at multiple interfaces in optical lens coatings. This approach, equivalent to Abeles' matrix method used in optical reflectometry, enables us to handle the complexity associated with inclusion of waves propagating outward that are created by reflection at boundaries within the layer structure. Matching of the tangential electric field vectors and the continuity requirements on the amplitude and gradient of the electric field at each interface, assumed for the moment to be abrupt, leads to

$$a_{l-1}E_{l-1}^{T} + \frac{E_{l-1}^{R}}{a_{l-1}} = \frac{E_{l}^{T}}{a_{1}} + a_{l}E_{l}^{R} \tag{8.27}$$

where E_l^R and E_l^T are, respectively, the reflected and transmitted components of the electric field in layer l and a_l is a phase factor related to the layer thickness, d_l, such that

$$a_l = \exp\left(-\frac{ikf_l d_l}{2}\right) \tag{8.28}$$

where f_l is the scattering factor for layer l expressed in terms of the absorption, dispersion, and incident angle, θ, such that:

$$f_l = \left(\theta_1^2 - 2\delta_l - 2i\beta_l\right)^{\frac{1}{2}} \tag{8.29}$$

The reflection amplitude coefficient $R_{l-1,l}$ between layers $l - 1$ and l, defined as

$$R_{l-1,l} = a_{l-1}^{2}\left[\frac{E_{l-1}^{R}}{E_{l-1}}\right] \tag{8.30}$$

can then be related to that at the interface between layers $l + 1$ and l. We obtain

$$R_{l-1,l} = a_{l-1}^{4}\left[\frac{R_{l,l+1} + F_{l-1,l}^{R}}{R_{l,l+1}F_{l-1,l}^{R} + 1}\right] \tag{8.31}$$

Here, $F_{l-1,l}^R$ is the Fresnel reflection coefficient between layer $l - 1$ and l:

$$F_{l-1,l}^{R} = \frac{\left(f_{l-1} - f_l\right)}{\left(f_{l-1} + f_l\right)} \tag{8.32}$$

Equation 8.32 is modified for the case where the interfaces are not abrupt, as in Equation 8.22, to

$$F_{l-1,l}^{R} = \frac{\left(f_{l-1} - f_l\right)}{\left(f_{l-1} + f_l\right)} \exp\left(\frac{-Q_z^l Q_z^{l-1} \sigma_l^2}{2}\right) \tag{8.33}$$

In a system of layers terminated by a semi-infinite substrate there is no reflected wave incident from below, and therefore $R_{l,l+1} = 0$ at the bottom of the stack. This, inserted into Equation 8.31, forms the starting point for an iterative calculation of Equation 8.31 for successive layers, working up to the surface. At the surface, we can then determine the ratio of the incident to reflected intensity from

$$\frac{I_R}{I_0} = \left|\frac{E_1^R}{E_1}\right|^2 = \left|R_{1,2}\right|^2 \tag{8.34}$$

Such formalism is ideally suited to numerical calculation and is remarkably fast on modern desktop or laptop computers. It forms the basis of almost all simulation of reflectivity curves. As we discussed elsewhere, because the phase information in Equation 8.30 is lost when the complex conjugate is taken in Equation 8.34 to reach the observed intensity, we cannot simply invert the scattered x-ray data to obtain the structure. Rather, we must simulate the scatter from a model structure and successively refine the model to obtain the best fit to the experimental data. Having a formalism that permits very rapid calculation of the reflectivity from a structure of an arbitrary number of layers is vital for the x-ray metrological method.

8.5.1 Reflectivity from a Bilayer

In the weak scattering limit, the scattered intensity is related to the Fourier transform of the layer structure. When more than one layer exists, each layer will contribute a period to the scattering profile corresponding to the inverse of the individual layer thickness. Thus, if there exists a thin layer of oxide on the surface of a metal, typically 1 nm thick, the reflectivity curve will be modulated by a long wavelength period. Because the intensity falls rapidly with scattering angle, only one or at best two of these fringes may be seen. Inclusion of the oxide layer in the model structure may nevertheless be crucial to obtaining a good fit between simulation and experiment.

When there are two layers of almost equal thickness, the highest-frequency period will correspond to the sum of the two layers, the individual layer thickness modulating the reflectivity at half the frequency of the bilayer period. Figure 8.11 shows an example of a GaAs/$Al_xGa_{1-x}As$ bilayer grown

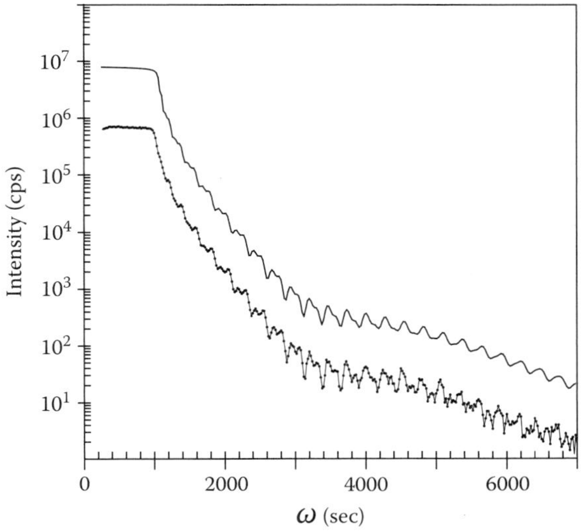

FIGURE 8.11
Experimental and fitted specular reflectivity from a GaAs/$Al_xGa_{1-x}As$ bilayer grown on a GaAs substrate. The $Al_{0.22}Ga_{0.78}As$ layer thickness was determined to be 50.5 ± 0.2 nm, and that of the top GaAs layer 52.1 ± 0.1 nm, with a 0.2 ± 0.1 nm oxide layer at the surface. Corresponding interface widths were 0.3 ± 0.3, 0.56 ± 0.015, and 1.00 ± 0.02 nm. The upper curve is the simulation, displaced up one decade for clarity.

on a GaAs substrate (previously shown in Chapter 2). The broad hump at high angle arises from a thin oxide layer.

For more than two or three layers, the matching of a model structure to the data, even using automatic fitting routines, can be extremely difficult. This is particularly true for layers where the electron density is very similar. It is important to have a nominal structure from the material producer; blind fitting can be very time-consuming.

8.5.2 Reflectivity from a Periodic Multilayer

Just as a periodic array of atoms within the crystal lattice results in Bragg peaks, so a periodic stack of layers of composition ABAB … results in low-angle Bragg peaks in the specular reflectivity. In exact analogy, the more repeats m in the stack $[AB]_m$, the sharper are the associated Bragg peaks. The positions of the Bragg peaks for a bilayer repeat of thicknesses d_A and d_B for the components A and B, respectively, are given by

$$n\lambda = 2\,(d_A + d_B)\,\sin\theta_I \tag{8.35}$$

Again, recalling that there is a frequency corresponding to each thickness, we find Kiessig fringes between the Bragg peaks with a period $\Delta\theta_I$ given by the total superlattice thickness

$$\Delta\theta_I = \lambda/[2m(d_A + d_B)] \tag{8.36}$$

Usually there are $m - 2$ fringes between the multilayer Bragg peaks. The relative intensity of successive Bragg peaks depends on the relative thickness d_A to d_B. Again, some insights may be obtained by remembering that the scattered amplitude is the Fourier transform of the real structure. For example, if $d_A = d_B$ we have a square wave, the odd harmonics of which are absent in the Fourier series. Thus, the even-order Bragg peaks, corresponding to the first, third, etc., harmonic, are absent. Similarly, if $d_A = 2d_B$, the third-order Bragg peak is absent. As the Parratt recursive formalism is also ideal for simulating multilayer structures, it is usual to allow autofitting routines to solve for the relative thickness. However, the above reasoning provides a useful check on the robustness of the model derived. Figure 8.12 shows an example of a periodic multilayer of 20 repeats of Au/Fe grown by molecular beam epitaxy on MgO (001). We note that the specular reflectivity falls slowly with wave vector and that there are only 17 fringes between the first and second Bragg peaks. The spacing and number of the Kiessig fringes between the Bragg peaks show that the bottom layer of the multilayer differs from the rest of the stack, and it was necessary to model this as an AuFe alloy layer to get a satisfactory fit to the data. The multilayer Bragg peak heights and overall fall in intensity with scattering angle are matched very well by simulation and give a best fit width of 0.21 ± 0.01 nm for the Fe-Au interface and 01.1 ± 0.01 nm for the Au-Fe interface.

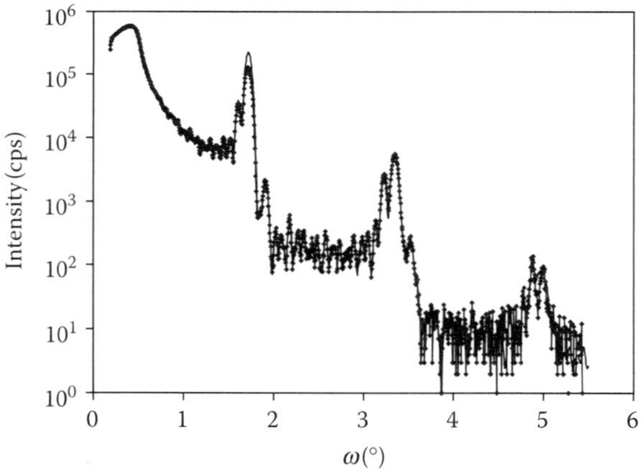

FIGURE 8.12
Specular reflectivity (marked with data points) and best-fit simulation of a 20-period AuFe multilayer grown on the cube face of MgO. (From A. Cole et al., *J. Phys. Condens. Matter.* 16, 1197–1209, 2004. With permission from IOP Publishing, Ltd.)

8.6 Summary

- Matching of electromagnetic waves at the boundary between materials of different optical density, as for visible light optics, provides all the theory necessary to interpret the grazing incidence specular reflectivity of x-rays from materials.
- Kiessig fringes, arising from interference between waves scattered from the bottom and top surfaces of thin films, provide a robust method for measurement of the thickness of nanometer-scale films with better than a 10th of a nanometer precision.
- As the method relies only on angular measurement and determination of the wavelength of an x-ray characteristic line, the method is an excellent and traceable metrological tool for use on thin films for which visible light methods fail due to the long wavelength of visible light.
- Inclusion of interface width into the models enables the total interface width due to a combination of topological roughness and compositional grading to be determined from the reflectivity.
- As seen in the next chapter, it is necessary to measure the diffuse scatter away from the specular condition in order to separate out these two components.

References

1. L. Névot and P. Croce, *Rev. Phys. Applique*, 15, 761, 1980.
2. M. Wormington, I. Pape, T.P.A. Hase, B.K. Tanner, and D.K. Bowen, *Phil. Mag. Lett.*, 74 (1996) 211–216.
3. B.K. Tanner, D.A. Allwood, and N.J. Mason, *Mater. Sci. Eng. B*, 80 (2001) 99–103.

9

X-ray Diffuse Scattering

9.1 Origin of Diffuse Scatter from Surfaces and Interfaces

As we saw in the last chapter, fitting the intensity of the specular x-ray scatter to a model layer structure enables the x-ray metrologist to measure the interface widths, in simple cases to better than 0.5 nm precision. However, the effects on the specular scatter of a compositionally graded interface and a topologically rough interface are identical. There is no information in the specular scatter about the structure in the plane of the surface or interface. If we model the variations for the averaged electron density across both these types of interface as error functions, we can add in quadrature the root mean squared (rms) roughness σ_t and the grading width Σ to give the total interface width σ. Explicitly this is

$$\sigma^2 = \sigma_t^2 + \Sigma^2 \tag{9.1}$$

The compositionally graded interface remains parallel to the surface and everywhere perpendicular to the scattering vector. All changes in the specular scatter due to the composition variation therefore must arise from coherent effects. On the contrary, the topologically rough interface is locally oriented in different directions and the reduction in specular scatter arises through diffuse scatter out of the specular direction. Measurement of this diffuse scatter and its variation in angular space provides a powerful method, complementary to scanning probe techniques, for the measurement of surface roughness and a unique method for the measurement of topological roughness of buried interfaces. By combination of specular and diffuse scatter measurement, we can distinguish between the three parameters in Equation 9.1, as the specular scatter measures σ and the diffuse scatter arises only from σ_t.

9.2 The Born Approximation

The case where the scattering is weak can be described in what is known as the Born wave approximation. Before embarking on a formal treatment, we will derive a simple expression for the integrated diffuse scatter that provides a rapid method of estimating the mean roughness by comparison with the integrated specular scatter. Let us consider a fixed value of the scattering vector component normal to the average surface Q_z, which just corresponds to keeping the detector fixed at a constant angle ($Q_z = 2\pi\Phi$, where Φ is the detector or, equivalently, scattering angle). We saw in Chapter 8 that the effect roughness on the Fresnel reflection *amplitude* was a reduction by a static Debye–Waller factor of the form

$$F^R_{layer} = F^R_{ideal-layer} \exp\left[-\tfrac{1}{2}\left(Q_z^2\sigma_t^2\right)\right] \tag{9.2}$$

The specularly reflected *intensity* I_{spec} therefore is reduced from the value for a perfectly smooth surface I_{ideal} by the square of this phase factor, giving

$$I_{spec} = I_{ideal} \exp\left[-Q_z^2\sigma_t^2\right] \tag{9.3}$$

Now the intensity difference ($I_{ideal} - I_{spec}$) must have been scattered into the diffuse scatter, through conservation of energy. Thus, the total diffuse scatter intensity I_{diff} is simply

$$I_{diff} = I_{ideal} - I_{spec} \tag{9.4}$$

Hence,

$$I_{diff} = I_{spec} \exp[Q_z^2\sigma_t^2] - I_{spec} \tag{9.5}$$

or

$$\frac{I_{diff}}{I_{spec}} = (\exp[Q_z^2\sigma_t^2]) - 1 \tag{9.6}$$

Equation 9.6 is immediately useful. For fixed detector position, we simply need to measure the total specular intensity and the total diffuse intensity. This can be done by scanning the specimen in what is conventionally known as a rocking curve. Figure 9.1 shows an example of the specular and diffuse scatter from two different layers of ZnTe on GaSb as a function of the

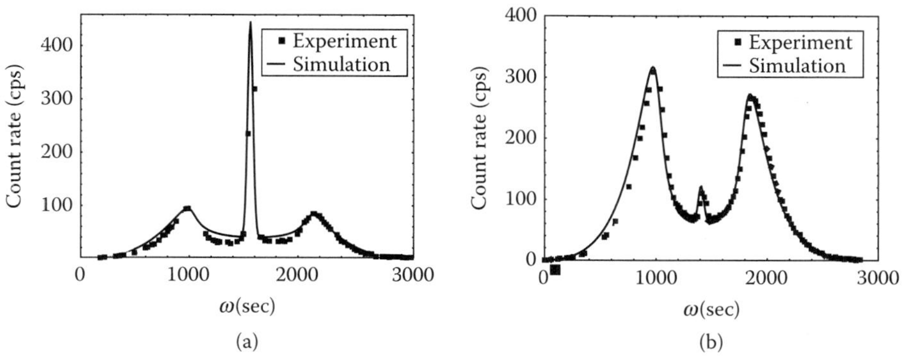

FIGURE 9.1
Rocking curves, corrected for beam footprint variation, at detector angle 2800 arc sec, for two ZnTe layers of different roughness (both grown on GaSb). Measurement of the relative areas under the specular peak in the center of each scan and the total diffuse scatter provides a measure of the roughness for (a) 2.0 nm and (b) 4.7 nm rms. (The origin of Yoneda wings in the diffuse scatter is discussed later.) The full simulations were done using the distorted-wave Born approximation theory described below. (Reused with permission from C.R. Li, *Journal of Applied Physics*, 82, 2281 (1997). Copyright 1997, American Institute of Physics.)

specimen angle. The specular peak is that in the center of each plot, and I_{spec} is obtained by measurement of the integral of the intensity under that peak. I_{diff} corresponds to the integral of the remaining scatter as a function of angle. In Figure 9.1b, the diffuse scatter is much greater than the specular scatter, but this is rarely the case for semiconductor surfaces.

There are three comments of caution regarding the quantitative use of Equation 9.6. The first is that the measurement is restricted to a range of angles between the critical angle for the incident beam and the critical angle for the exit beam. Particularly for interfaces and surfaces that are rough on a short-length scale, the scattering is widely distributed in angular space and only a fraction of the diffusely scattered intensity is collected in the rocking curve scan. The roughness is therefore underestimated. (Note, however, that the effect of this limited collection range is properly treated in simulation programs.) Second, the roughness determined is a weighted sum of roughness from all the interfaces, weighted over the electron density difference between various interfaces. For multiple-layer structures, detailed interpretation is impossible. Third, for very rough interfaces, the approximation of simple scattering may break down as the diffusely scattered waves are rescattered. Equation 9.6 blows up as σ_t increases and becomes invalid for large values. While it also fails at large detector angle (scattering vector Q_z), this is rarely a problem as the intensity of the specular scatter itself decreases rapidly with Q_z (see Chapter 8) and measurement of the specular intensity becomes impossible. The Born approximation also fails to predict the Yoneda wings, enhanced diffuse scatter that arises when the incident or exit beam satisfies the critical angle.

Let us now examine the Born approximation formally. We make the assumption that the scattering from the material comes from point-like objects with the scattering potential $V(r)$. Within the medium the potential is

$$V = k_0^2\left(1 - n^2\right) \tag{9.7}$$

where n is the refractive index of the material and k_0 $(= 2\pi/\lambda)$ the magnitude of the wave vector (wave number) of the electromagnetic wave in vacuum. The variation of the scattered intensity as a function of solid angle Ω in space can be described by a cross section s that describes the amount of scatter in terms of an equivalent area of a fully opaque obstacle. Using simple kinematical scattering theory, we can express the differential of the scattering cross section as

$$\frac{ds}{d\Omega} = N^2 r_e^{\,2} \int_V dr \int_V \exp\left[-i\mathbf{Q}\cdot\left(\mathbf{r} - \mathbf{r}'\right)\right] dr' \tag{9.8}$$

where N is the number density of the particles involved in the scattering and r_e is the classical electron radius defined as

$$r_e = e^2 / 4\pi\varepsilon_0 mc^2 \tag{9.9}$$

with m the electron mass, e the electronic charge, and c the velocity of light. We will revisit these parameters in the next chapter.

Equation 9.8 can be transformed into a surface integral over a surface S_0 by the application of Stokes' theorem, such that

$$\frac{ds}{d\Omega} = \frac{N^2 r_e^2}{Q_z^2} \iint_{S_0} dxdy \iint_{S_0} \exp\left(-iq_z\left[z(x,y) - z(x',y')\right]\right) \exp\left(-i\left[Q_x\left(x - x'\right) + Q_y\left(y - y'\right)\right]\right) dx'dy' \tag{9.10}$$

where the vector, $\mathbf{r} = \mathbf{r} - \mathbf{r}'$, in the first equation is now expressed explicitly in terms of the in-plane, (x, y), and out-of-plane, $z(x, y)$, components of direction. The key now is to describe the height–height variation statistically. If we treat the out-of-plane component as a Gaussian random variable, we have

$$\left\langle \left[z(x',y') - z(x,y)\right]^2 \right\rangle = g\left(x' - x, y' - y\right) = g(R) \tag{9.11}$$

where $R = (X^2 + Y^2)^{1/2}$ and $X = x' - x$, $Y = y' - y$. The differential cross section can then be expressed in terms of the area, L_xL_y, illuminated by the incident radiation, giving

$$\frac{ds}{d\Omega} = \frac{N^2 r_e^2}{Q_z^2} L_x L_y \iint_{S_0} dX dY \exp\left[-\left(q_z^2 g(R)/2\right)\right] \exp\left[-i\left(Q_x X + Q_y Y\right)\right] \quad (9.12)$$

In this form it is possible to obtain explicit expressions for the cross section for different models of the height difference function, $g(R)$.

9.2.1 Interface Modeling within the Born Approximation

There have been a number of suggested functions $g(R)$ put forward to describe realistic interface models. One such quite widely used function has an exponential fall in the height difference function, $g(R)$, as a function of distance, namely,

$$g(R) = 2\sigma_t^2 \left[1 - \exp\left\{-\left(\frac{R}{\xi}\right)\right\}\right] \quad (9.13)$$

where the length ξ is called the lateral correlation length.

An important development occurred in 1988 when Sinha et al.[1] proposed the use of a self-affine fractal description. This incorporates a power law dependence through a fractal, or Hurst, parameter h, where $0 < h < 1$. $h = 1$ corresponds to a two-dimensional surface, while $h = 0$ corresponds to a three-dimensional interface. Examples of sections of surfaces with $h = 1$ and $h = 0.25$ are shown in Figure 9.2. The fractal parameter is thus related to the fractal dimension D by

$$D = 3 - h \quad (9.14)$$

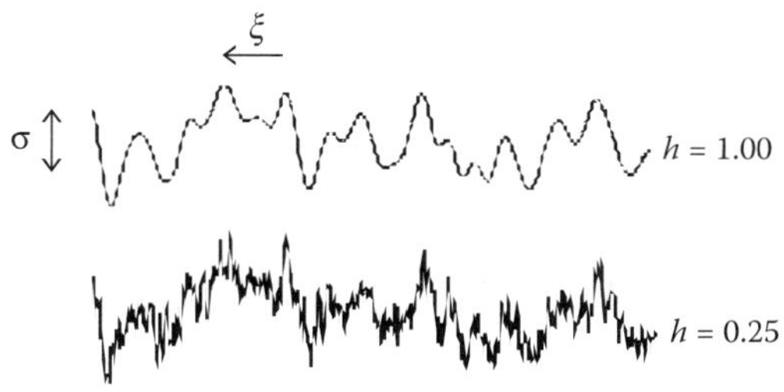

FIGURE 9.2
Simulated surface realizations (or interface profiles) with Hurst parameters h of (a) 1 and (b) 0.25.

The problem is to avoid such a power law relation blowing up at large distances. The solution is to introduce a cutoff at the length scale ξ. Such a cutoff power law dependence is satisfied for a height difference function of the form

$$g(R) = 2\sigma_t^2 \left[1 - \exp\left\{ -\left(\frac{R}{\xi} \right)^{2h} \right\} \right] \quad (9.15)$$

As the distance, R, tends toward infinity, the relative height between two points described by the height difference function tends to $2\sigma^2$.

In modeling the diffuse scatter from interfaces it is preferable to consider a height–height correlation function, $C(R)$, rather than the height difference function. This correlation function provides a measure of the degree of knowledge between two points on a surface, separated by a distance R. The height–height correlation function is defined as

$$C(R) = \langle z(R) z(0,0) \rangle = \sigma^2 - \frac{1}{2} g(R) \quad (9.16)$$

which can then be expressed, via Equation 9.15, as

$$C(R) = \sigma^2 \exp\left[-\left(\frac{R}{\xi} \right)^{2h} \right] \quad (9.17)$$

Figure 9.3 shows this height–height correlation function as a function of distance R for several values of the fractal parameter h. The common point of the curves corresponds to the length of the surface vector that is equal to the length scale in the surface over which $C(R)$ falls to $1/e$ of its value at $R = 0$. This length scale, 20 nm in the case illustrated, is the lateral correlation length, ξ, and defines the length scale below which the surface is fractal in nature.

Equation 9.12, the differential cross section, becomes

$$\frac{ds}{d\Omega} = \frac{N^2 r_e^2}{Q_z^2} L_x L_y \exp\left[-Q_z^2 \sigma_t^2 \right] \iint_{S_0} dX dY \exp\left[Q_z^2 C(R) \right] \exp\left[-i\left(Q_x X + Q_y Y \right) \right] \quad (9.18)$$

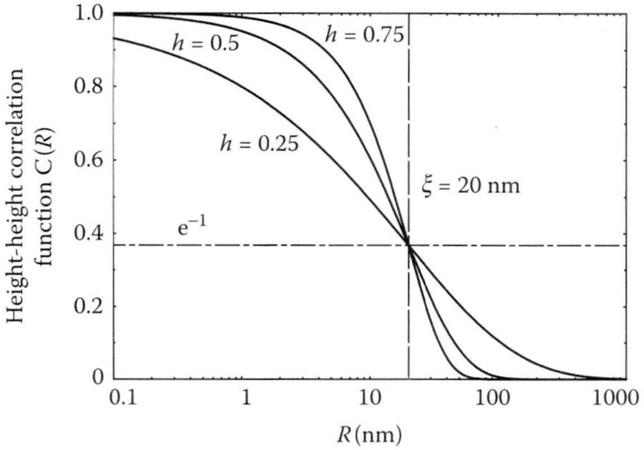

FIGURE 9.3
Height–height correlation function as a function of distance in the interface plane R for various values of h. All curves have $\xi = 20$ nm.

and it is possible to subdivide the terms in this expression so as to express the specular and diffuse components separately. Explicitly, these are

$$s_{spec}\left(Q\right) = \left| F_l^R \right|^2 \exp\left(-Q_z^2\sigma_t^2\right) \quad (9.19)$$

$$s_{diff}\left(Q\right) = \frac{2\pi}{Q_z^2} \exp\left(-Q_z^2\sigma_t^2\right) \int_0^{\infty} R\left[\exp\left(Q_z^2\sigma_t^2 e^{-\left(iR/\xi\right)^{2h}}\right) - 1\right] J_0\left(Q_{x,y}R\right) dR \quad (9.20)$$

where $s(Q)N^2r_e^2$ is the differential cross section per unit area of sample, F_l^R is the Fresnel amplitude reflection coefficient, and $J_0(Q_{x,y}R)$ is a Bessel function of the first kind. Equation 9.19 is equivalent to Equation 9.3.

9.3 The Distorted-Wave Born Approximation

Within the Born approximation, the rough surface as a whole is taken to be the perturbation potential on the propagation of the incident plane wave. This is adequate for very weak scattering and small roughness of that surface. However, a much better approach is to assume that the difference between a rough surface and an ideal surface, located at the same average position, acts as a perturbation. In the distorted-wave Born approximation (DWBA) theory, the perturbation associated with the rough interface is applied to the exact solutions to the wave equation associated with an ideal interface (Fresnel waves). This is done by separating the scattering potential into two distinct terms:

$$V = V_1 + V_2 \tag{9.21}$$

where V_1 is the potential of the ideal system and V_2 is the potential that acts as a disturbance to this ideal case. For an interface at average position z = 0, we can formally describe the rough surface potential as

$$V_1 = \begin{cases} k_0^2\left(1-n^2\right) & -a < z < 0 \\ 0 & z > 0 \end{cases} \tag{9.22}$$

and

$$V_2 = \begin{cases} k_0^2\left(1-n^2\right) & 0 < z < z(x,y) \ \text{ if } z(x,y) > 0 \\ -k_0^2\left(1-n^2\right) & z(x,y) < z < 0 \ \text{ if } z(x,y) < 0 \\ 0 & \text{elsewhere} \end{cases} \tag{9.23}$$

in which $z(x, y)$ represents the statistical average of the surface. The reflected and transmitted plane waves generated at the ideal interface experience the perturbation generated by the roughness potential, V_2.

The expression for the specular scatter, when the roughness is described by Gaussian statistics, is given by the same modification of the Fresnel coefficient as Névot and Croce,[3] derived (Equation 8.22) by a different method, namely,

$$\left|R\left(k\right)\right|^2 = |F_l^R|^2 \exp\left(-Q_z Q_z^t \sigma_t^2\right) \tag{9.24}$$

After much mathematical manipulation, to be found in the seminal paper of Sinha et al.,[1] the differential cross section for the diffuse scatter can be shown to be

$$\left[\frac{d\sigma}{d\Omega}\right]_{diffuse} = L_x L_y \frac{\left|k_0^2\left(1-n^2\right)\right|^2}{16\pi^2} \left|T\left(k_1\right)\right|^2 \left|T\left(k_2\right)\right|^2 s\left(Q_t\right) \tag{9.25}$$

where $|T(k_m)|^2$ are the Fresnel transmission coefficients for the incident (m = 1) and scattered (m = 2) waves, including (in analogy to Equation 9.24) the exponential damping term, into which is incorporated the roughness, and $L_X L_Y$ defines the illuminated area. The $s(Q_t)$ term contains a Fourier transform that must be calculated numerically.

We can now see the origin of the wings in the diffuse scatter noted in Figure 9.1 and originally reported by Yoneda. These occur at angles where the incident and exit waves, respectively, make the critical angle θ_c with the surface. At these points there is a maximum in $T(k_m)$ as $k_{T,Z}$ goes to zero in Equation 8.9, resulting in a maximum in the diffuse scatter at the critical angle. We also see that the peaks are symmetric about the specular peak, Equation 9.25 being a product of the Fresnel transmissivities for the incident and exit beams. Another way of understanding the origin of the symmetrically located peaks is to note that at the critical angle, the energy flow is in the direction of the surface and the electric field at the surface is doubled in amplitude (hence quadrupled in intensity). The propagation of energy within the surface results in enhanced diffuse scatter from topological roughness. It is not so easy to see why the diffuse scatter peaks at the critical angle associated with the exit wave making the critical angle to the surface. However, there is an important theorem in optics, called the reciprocity theorem, which states that if we reverse the directions of the rays, we must have the same results. Thus, in the case of the exit wave grazing the surface on reversal, it becomes the incident wave generating enhanced scatter.

The position of the Yoneda wings is an excellent metrological technique for determination of the critical angle. Indeed, the peak maximum can be located much more precisely than the critical angle in the specular reflectivity, where the task is to locate the point at which the intensity starts to fall rapidly. In favorable cases, for example, in Figure 9.1, the peak is very sharp and well defined. As the critical angle is dependent on the near-surface density (Figure 9.4), porosity or grain pull-out will reduce the value of the critical angle below that of the bulk material, and measurement of the Yoneda peak position provides a method for this determination. It is important to note at this point that the asymmetry in the intensity of the Yoneda wings has no fundamental significance. The high-angle wing is always less intense than the low-angle wing because the beam footprint on the sample is less at the high angle. For a beam of width w, the footprint on a large sample is $w/\sin\alpha$, where α is in the angle of the incident beam with the surface. Correction is not always straightforward, but division of the angular coordinate by the sine of the incidence angle α usually produces roughly equal peak heights. The footprint correction is built into commercial software packages for simulation of the diffuse scatter.

9.3.1 Separation of Topological Roughness and Compositional Grading within the DWBA

We have noted several times now that, in the specular reflectivity, we cannot distinguish between a rough surface and one that is perfectly flat but has a

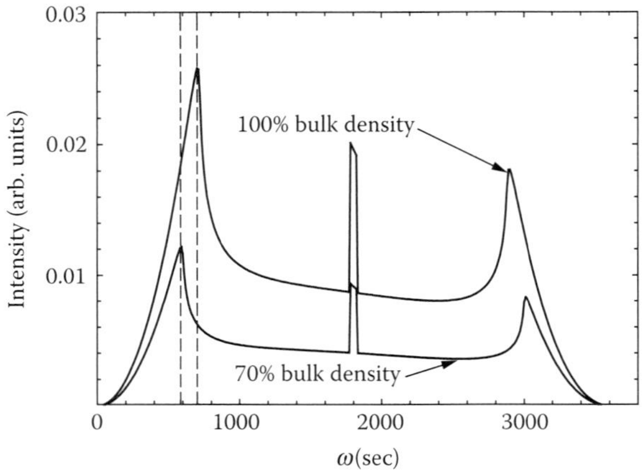

FIGURE 9.4
Effect of near-surface density on the position of the Yoneda wings in the diffuse scatter.

composition or density that varies with depth. In the case where both effects are present, the experimental data have less diffuse scatter than expected from the width of the interface determined from the specular reflectivity.

The case where grading and roughness both occur was first modeled by Wormington et al.,[2] who extended Sinha's theory to include a variation of the electron density perpendicular to the surface. This results in the following expression for the specular intensity:

$$I_s = I_0 \left| R(k_{1z}) e^{-Q_z Q_z^t \sigma_t^2/2} \int_{-\infty}^{\infty} [\mathrm{d}\rho(z)/\mathrm{d}z] e^{i\sqrt{Q_z Q_z^t}\, z} \mathrm{d}z \right|^2 \tag{9.26}$$

where I_0 is the incident intensity. The function $R(k_{1z})$ is the Fresnel reflection amplitude for an ideal surface and k_{1z} is the z component of the incident wave vector. The quantities Q_z and Q_z^t are the z components of the scattering vector in air and inside the scattering material, respectively. The coordinate system remains, with z directed perpendicular to the surface and x directed parallel to the surface. The rms roughness and electron density of the surface are denoted by σ_t and $\rho(z)$, respectively. It follows that the expression for the diffuse intensity is given by

$$\begin{aligned} I_d = I_0 \frac{k_0^3 \delta\theta_2}{8\pi \sin(\theta_1)} & \left| T(k_{1z}) T(k_{2z})(1-n^2) \int_{-\infty}^{\infty} [d\rho(z)/dz] e^{iQ_z^t z} dz \right|^2 \\ & \times \frac{e^{-[(Q_z^t)^2 + (Q_z^{t*})^2]\sigma_t^2/2}}{|Q_z^t|^2} \int_0^{\infty} (e^{|Q_z^t|^2 C(X)} - 1) \cos(Q_X X) dX \end{aligned} \tag{9.27}$$

where $T(k_{1z})$ and $T(k_{2z})$ are the Fresnel transmission amplitudes for an ideal surface and k_{2z} is the z component of the scattered wave vector. Here k_0 continues to correspond to the wave number of the incident x-rays and n is the refractive index deep within the scattering material. The x component of the scattering vector is given by Q_X, and the angular acceptance of the detector slit in the scattering plane is denoted by $\delta\theta_2$. The sine of the angle of incidence θ_1, which appears in the denominator of Equation 9.27, accounts for the change in the area illuminated by the x-ray beam as the sample is rotated.

Description of the interface is exactly as in the Born approximation. We must choose a method of statistically modeling the interface. Wormington et al.[2] used the three-parameter autocovariance function describing a self-affine fractal, introduced by Sinha et al.[1] This is

$$C(X) = \langle \eta(0)\eta(X) \rangle = \sigma_t^2 \exp(-|X/\xi|^{2h}) \tag{9.28}$$

Here, the random quantity $\eta(X)$ denotes the local position of the center of the electron density profile, X is the separation between two points in the plane of the surface, and $\langle \cdots \rangle$ denotes a configurational average over all surface points. The right-hand side of Equation 9.28 describes the behavior of a self-affine fractal surface with a cutoff determined by the correlation length ξ. Although this represents a particular class of surfaces, it is quite general in that it can describe both jagged and smoothly undulating rough surfaces, depending upon the value of the Hurst parameter, h.

The Fourier cosine integral in Equation 9.27 does not in general have an analytical solution. However, using this autocovariance function, it is possible to use a numerical approach, based upon lookup tables, to evaluate the integral rapidly and to a high degree of accuracy. If an error function of width Σ is used to model the intrinsic electron density grading, the Fourier integrals in Equations 9.26 and 9.27 become $\exp(-Q_z Q_z^t \Sigma^2/2)$ and $\exp(-Q_z^t \Sigma^2/2)$, respectively. These factors reduce the specular and diffuse intensity as the scattering angle increases and explain the apparent decrease in the roughness required to fit, with the Sinha model, the diffuse scatter in rocking curves taken at successively higher detector angles.

Table 9.1 shows the roughness parameters determined by Wormington et al.[2] from two polished samples (A and B) of the low-thermal-expansion oxide ceramic glass Zerodur®. Equations 9.26 to 9.28 were used to fit the data for the specular scatter and the diffuse scatter in one rocking curve. The other two diffuse scatter curves shown in Figure 9.5 contained no adjustable parameter. Agreement is excellent (as was the case with two additional curves never published). A further interesting point is that the correlation length ξ is comparable with the particle size of the polishing paste and that addition of σ_t and Σ in quadrature results in values for the interface widths in good agreement with the values of 1.18 ± 0.03 and 1.40 ± 0.03 nm for the two samples A and B, deduced from the specular reflectivity.

TABLE 9.1

Roughness Parameters Derived from Speular and Diffuse Scatter from Two Samples of Polished Zerodur(R)

Sample	σ_t (nm)	Σ (nm)	ξ (nm)	h	$\sqrt{\sigma_\tau^2 + \Sigma^2}$ (nm)
A	1.03 ± 0.05	0.62 ± 0.06	900 ± 100	0.30 ± 0.03	1.20 ± 0.08
B	1.28 ± 0.03	0.68 ± 0.05	1300 ± 200	0.30 ± 0.03	1.44 ± 0.08

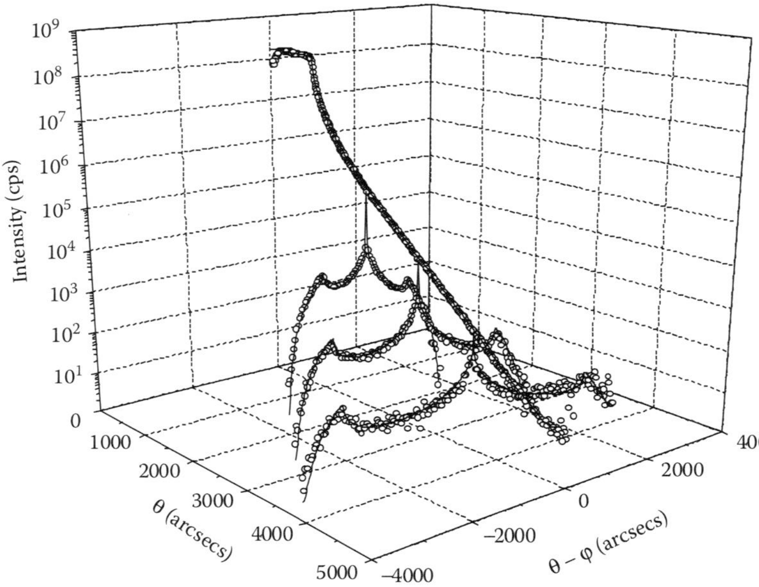

FIGURE 9.5
Experimental and simulated specular and diffuse scatter from a polished Zerodur® sample. One set of parameters fits all the data.

9.4 Effect of Interface Parameters on Diffuse Scatter

The effect on the diffuse scatter of increase in the roughness amplitude σ_t is, as might be deduced from the beginning of Section 9.2, to scale the diffuse scatter with respect to the specular scatter. Changing the rms roughness from 1 to 0.4 nm simply results in a scaling of the diffuse scatter curve (Figure 9.6), that is, a displacement on a logarithmic scale.

The effects of changing correlation length ξ and Hurst fractal parameter h are not always uncoupled. As a general rule, for long correlation length and

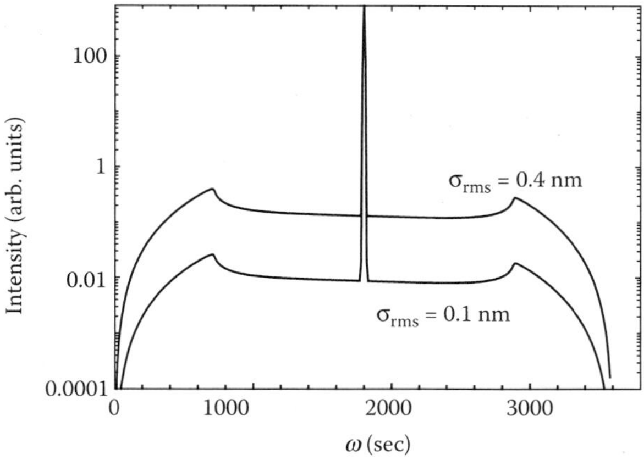

FIGURE 9.6
Effect of increasing rms roughness on the diffuse scatter distribution.

large h, the diffuse scatter appears close to the specular peak. This reflects the fact that scattering from long-length scale features appears at a small scattering angle. Conversely, short ξ and small h result in scatter widely distributed in space. Examples of the difference in diffuse scatter distribution between short and long ξ are given in Figure 9.7. The effect of varying h is shown in Figure 9.8. Again, however, it should be emphasized that ξ and h have very similar effects for some combinations of values. The difference between a surface with large h but very short correlation length and one with small h but long correlation length is not obvious. Simulation and fitting of the diffuse scattering intensity distribution is required to investigate the detailed in-plane structure of surfaces and interfaces.

9.5 Multiple-Layer Structures

We have already indicated that the diffuse scatter arises from scattering at all interfaces and that the contribution to the scatter depends on the electron density difference between the interfaces. Usually, the top surface roughness, where the electron density difference is between zero at the vacuum (air) interface and the material, gives rise to the most significant scatter. It is thus very difficult to separate the effects of individual interfaces. One possible method lies in the use of a period multilayer, when one type of interface roughness appears in a rather special angular position.

In a multiple-layer structure, the roughness on one layer may be conformal with the roughness of the layer below it, or it may have a random relation with respect to it. There can thus be defined two types of roughness, conformal (or correlated) roughness σ_c (Figure 9.9a) and random (or uncorrelated)

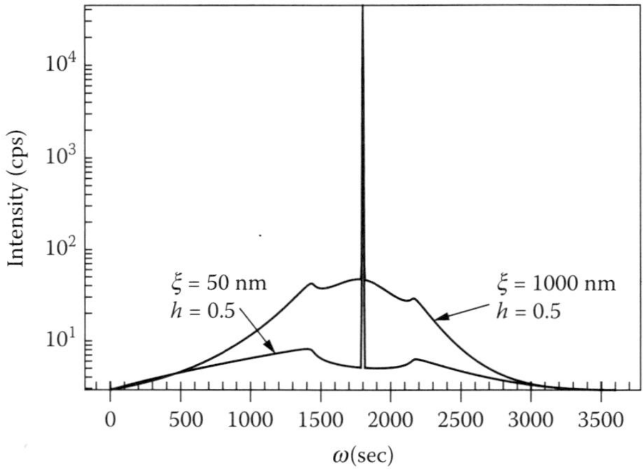

FIGURE 9.7
Examples of the difference in diffuse scatter distribution between short and long ξ.

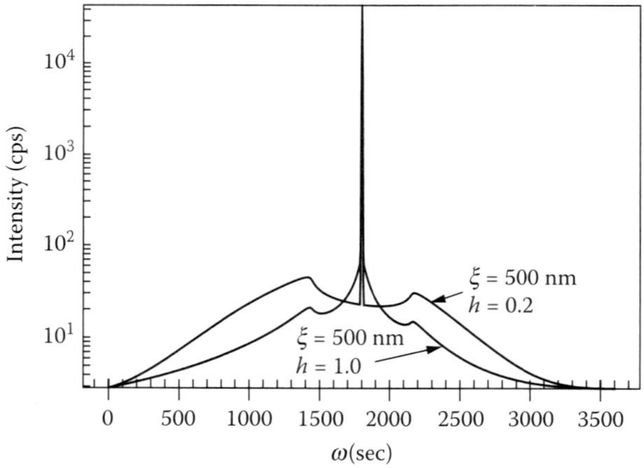

FIGURE 9.8
The effect of varying h.

roughness σ_u (Figure 9.9b). If we assume a Gaussian roughness, the total roughness, σ_t, is the sum in quadrature of the two components:

$$\sigma_t^2 = \sigma_c^2 + \sigma_u^2 \tag{9.29}$$

Holý and colleagues[4] extended the model to include the out-of-plane correlation by use of a covariance function of the form

$$C_{j,k}(R) = \langle \delta z_j(0) \delta z_k(R) \rangle \tag{9.30}$$

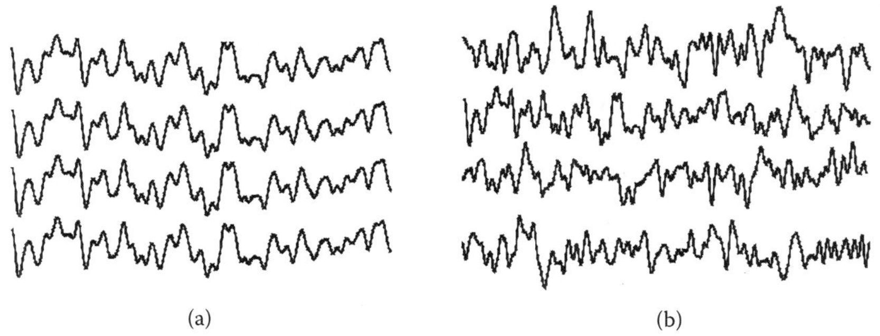

FIGURE 9.9
Computer-generated interfaces showing (a) conformal and (b) random roughness.

where δz_j and δz_k refer to the local centers of the roughness profile at the jth and kth interfaces. This has been built into two models.

The first is to define an out-of-plane correlation length ζ as a length scale over which replication of the roughness extends through the layer stack. Explicitly, it is defined as the out-of-plane distance over which the correlations between the jth and kth interfaces are damped by a factor of e^{-1}. Using this model, the covariance function becomes:

$$C_{j,k}(R) = \sigma_{j,k}^2 \exp\left[-\left(\frac{R}{\xi_{j,k}}\right)^{2h}\right] \exp\left[\frac{-|z_j - z_k|}{\zeta}\right] \quad (9.31)$$

where

$$\sigma_{j,k}^2 = \sigma_j^2 + \sigma_k^2 \quad \text{and} \quad \xi_{j,k}^{2h} = \frac{\xi_j^{2h} + \xi_k^{2h}}{2}$$

While this model is physically very reasonable, and the roughness amplitude and correlation length can be specified for each layer, computation time is long for systems with more than a few layers, as the calculation time scales with the square of the number of layers.

A model whose computation time scales linearly with layer number, but is less physically realistic, is one in which there is specified a constant fraction of correlated and uncorrelated roughness at each interface. This fraction does not change through the multilayer, and similarly, the lateral correlation length and fractal parameter are common for all interfaces. The covariance function for this model is

$$C_{j,k}(R) = \left(\sigma_{u,j}\sigma_{u,k}\delta_{j,k} + \sigma_{c,j}\sigma_{c,k}\right)\exp\left[-\left(\frac{R}{\xi}\right)^{2h}\right] \quad (9.32)$$

The roughness at each interface is represented by an intrinsic component, $\sigma_{u,j}$, and a component that has originated from the substrate and replicated upwards exactly, the $\sigma_{c,j}$ term.

The importance of the conformal roughness is that because of its coherence, the diffuse scatter from successive layers with this type of roughness is also coherent. Thus, interference effects are seen in the diffuse scatter. In the case of a periodic multilayer, strong scattering occurs at the scattering angle corresponding to the Bragg condition. The most important type of instrument scan for revealing these interference effects is a coupled specimen and detector scan in which the detector moves at twice the rate of the specimen. It differs from the specular ($\omega - 2\theta$) scan in that the specimen is initially offset by a small amount, which must be greater than the angular width of the specular peak, from the zero position. The diffuse scatter measured is thus close to the specular scatter. In measuring the specular reflectivity, this forward diffuse scatter must be subtracted from the measured specular scatter to obtain the true specular reflectivity. Such a subtraction is particularly important for surfaces and interfaces that have high roughness amplitude, because what often appears at first sight to be specular scatter will be diffuse scatter in the forward direction. This is particularly true for large values of scattering angle (vector).

Figure 9.10 is an example of the specular and offset coupled scans from a thin layer of Au grown by molecular beam epitaxy on a 3-nm Fe buffer layer on (001) MgO. The high-frequency fringes in the specular scatter, corresponding to the layer thickness, are replicated in the interference fringes in the

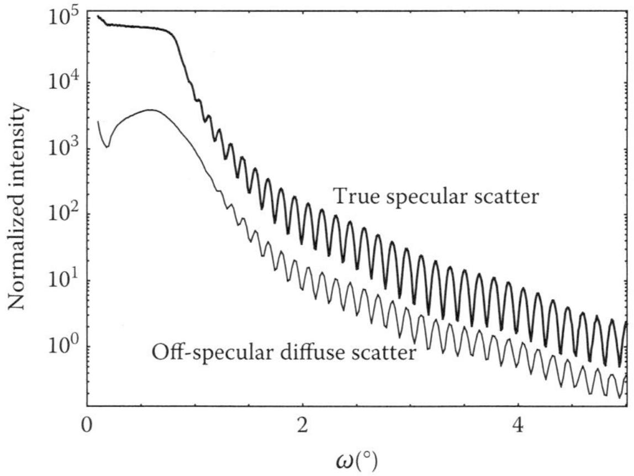

FIGURE 9.10
Specular and off-specular coupled specimen (one unit) and detector (two units) scans of the scatter from a 49.4 (± 0.1)-nm epitaxial layer of Au, surface rms roughness of 0.1 nm, grown by MBE on a 0.6-nm Fe buffer layer on a 0.5-nm roughness (001) MgO substrate. The very high contrast interference fringes in the diffuse scatter indicate almost total conformality in the top and bottom surface roughness. (This film had a very long in-plane correlation length.) (From Cole, A. et al., *J. Phys. Condens. Matter*, 16, 1197–1209, 2004. With permission from IOP Publishing, Ltd.)

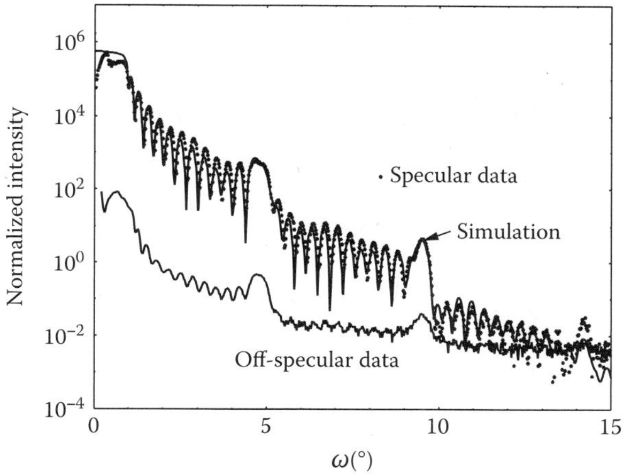

FIGURE 9.11
Specular and off-specular $\omega - 2\theta$ scans from a $[Co(0.3\ nm)/Pt(1.25\ nm)]_{10}$ multilayer grown on Si. (Courtesy A.S.H. Rozatian, Ph.D. thesis, Durham University, 2004. With permission.)

diffuse scatter. We can immediately deduce that the roughness of the substrate has replicated through the whole layer stack.

For a periodic multilayer, the coherence in the diffuse scatter can give rise to the off-specular Bragg peaks. An example of a $[Co(0.3\ nm)/Pt(1.25\ nm)]_{10}$ multilayer, grown on Si by magnetron sputtering, is shown in Figure 9.11. The replication of the roughness profile between successive layers gives rise to coherence in the diffuse scatter, and hence the off-specular Bragg peaks. The presence of the off-specular Kiessig fringes arises from the conformality of the top and bottom surfaces of the multilayer stack.

Figure 9.12 shows a series of off-specular simulations, using the covariance function of Equation 9.31, for a Fe/Au multilayer of total stack thickness 35 nm and strongly conformal roughness. When the out-of-plane correlation length, ζ, is in excess of the total stack thickness, all of the interfaces are highly correlated and the off-specular scan exhibits Kiessig fringes and Bragg peaks. When the out-of-plane correlation length becomes less than the stack thickness, the off-specular Kiessig fringes are lost, as the scatter originating at the substrate and top surface is no longer coherent. The Bragg peaks broaden due to the reduction in the number of correlated bilayers.

9.6 Diffuse Scatter Represented in Reciprocal Space

A very powerful way to represent the diffuse scatter is to plot it in reciprocal space. As we will see, reciprocal space does map quite straightforwardly

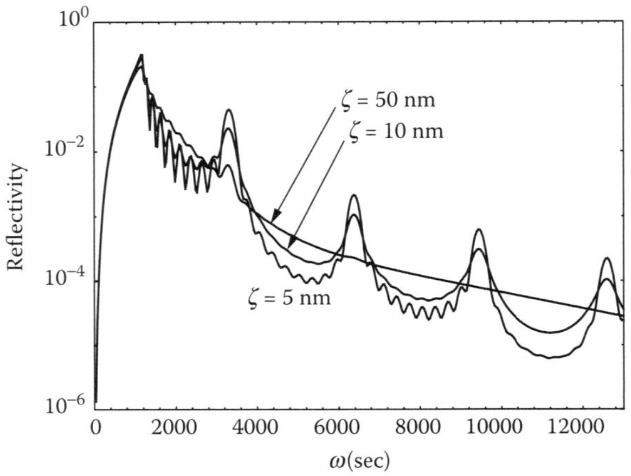

FIGURE 9.12
Simulated off-specular scans, with –0.1° offset, for a Fe/Au multilayer for different values of out-of-plane correlation length, ζ. (Courtesy B.D. Fulthorpe.)

into angular space, and it will give us a crucial tool for understanding diffraction from a crystalline material in Chapter 10.

We begin by recalling Equation 8.21 in the previous chapter. This definition of the scattering vector **Q** is represented graphically in Figure 9.13. We already noted in Figure 8.1 that, for the specular scattering, **Q** is oriented along the surface normal to the specimen surface. In the general case, **Q** can be oriented in other directions in the scattering plane. It is always zero if the scattering angle is zero, that is, the incident and scattered beams are in the same direction. As the scattering angle Φ (= $\alpha + \beta$) increases, so the length of Q increases. (Remember that $|\mathbf{k}_I| = |\mathbf{k}_R| = 2\pi/\lambda$ and cannot vary.)

The range of possible values of $Q(x, z)$ as a function of its x component (in the surface) and z component (normal to the surface) is shown in Figure 9.14. There are two shaded areas that are inaccessible because the incident or exit beams do not emerge from the surface. The rest of the area shown can be accessed as, for the low-angle scattering angles considered here, $|\mathbf{k}_I| >> |\mathbf{Q}|$. We can map the scattering in reciprocal space, so-called reciprocal space mapping, by performing a series of scans along lines in reciprocal space and combining them into a contour (or pseudocolor) map.

There are four commonly used scans encountered in grazing incidence x-ray reflectivity measurements.

9.6.1 Specular Scan

This is a coupled scan in which the specimen is incremented one unit of angle and the detector two units of angle at each step. Both are parallel to the specimen surface. Such a sequence means that **Q** starts at zero, and because the incident and scattered beams are always symmetric with respect to the surface, it extends along the normal to the surface, vertically in the diagram.

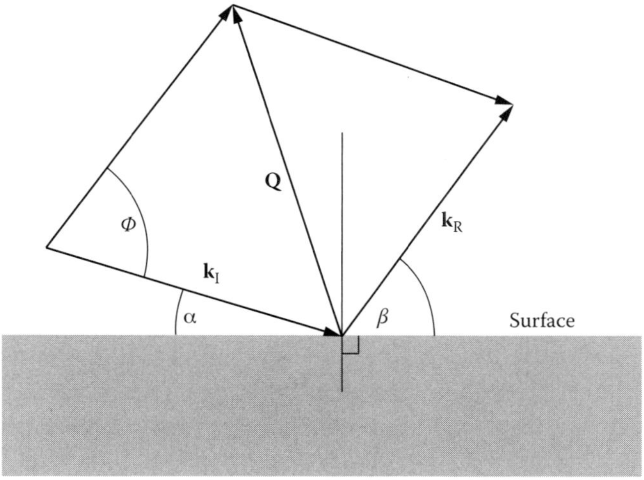

FIGURE 9.13
Definition of scattering vector **Q** and its relation to the specimen surface and angles of incidence and scattering.

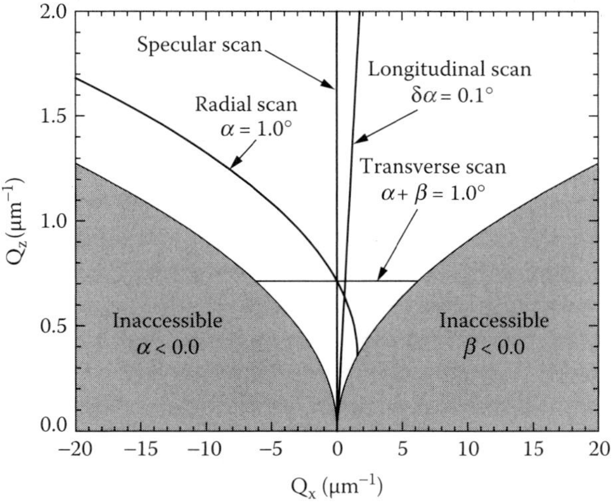

FIGURE 9.14
Accessible area of scattering in reciprocal space, showing the four key scans used for data collection.

9.6.2 Off-Specular Coupled Scan

This is again a coupled scan in which the specimen is incremented one unit of angle and the detector two units of angle at each step. However, although the incident beam and that collected in the detector aperture are parallel at the start of the scan, and hence $|Q| = 0$ initially, the sample is offset from the symmetric specular condition. Thus, **Q** traces out a straight line as for the specular reflection, but inclined at an angle to the surface-normal direction (or Q_z). Thus, immediately we see that we are probing the scatter with a value of $Q_x \neq 0$. Note that although, due to the scale of the diagram, the scan appears to give measurements for all angles along the scan line, there is always a region of inaccessibility when either the incident or exit beam does not emerge from the sample.

9.6.3 Transverse Scan

If the scattering angle Φ is kept fixed and the specimen only is scanned between the condition for grazing incidence and grazing exit, **Q** traces out an arc of a circle. However, on substitution of numbers, we find that the allowed values of Q_z are large compared with those of Q_x. The scale is compressed in the x direction. Therefore, the scan is, to a very good approximation, a straight line parallel to Q_x.

9.6.4 Radial Scan

The final type of scan has an important property that the incidence angle remains constant and the depth to which the incident beam probes into the sample remains constant. Here, the detector only is scanned. There is a region of inaccessibility. It is often used for porosity measurement (Chapter 7).

An example of a simulated reciprocal space map of a Fe/Au multilayer is shown in Figure 9.15. The structure is {Fe(1.5 nm)/Au(2 nm)} × 15 and is simulated using the covariance function of Equation 9.32, and for which the roughness of each layer was set at 0.6 nm, and the lateral correlation length, ξ, and fractal parameter, h, used were 20 and 0.25 nm, respectively. The roughness is totally conformal. There is asymmetry in the scatter about $Q_x = 0$ due to the inclusion of the variation in beam footprint with sample angle, and the very intense specular ridge running vertically upward has not been included.

For conformal roughness, the diffuse scatter is distributed in a very distinct manner in reciprocal space. There are strong bands of scatter, somewhat confusingly termed resonant diffuse sheets (RDS), at the positions of the Bragg peaks, arising from coherent scatter within the bilayers. The upward curvature of these diffuse sheets, picturesquely known as "Holý bananas" after Vaclav Holý, who first identified them, arises as a result of refraction. Between these sheets, corresponding to the periodicity of the multilayer, are bands of enhanced diffuse scatter corresponding to the Kiessig fringes in the off-specular scans. The spacing is determined by the total thickness of the multilayer. The continuous curved lines mark the positions of Bragg-like peaks and arise due to dynamical effects, discussed by Holý and Baumbach.[4] In the case of roughness that is totally random between layers, there is very little structure in the simulated map of the scattering; the intensity falls off monotonically with increasing value of Q.

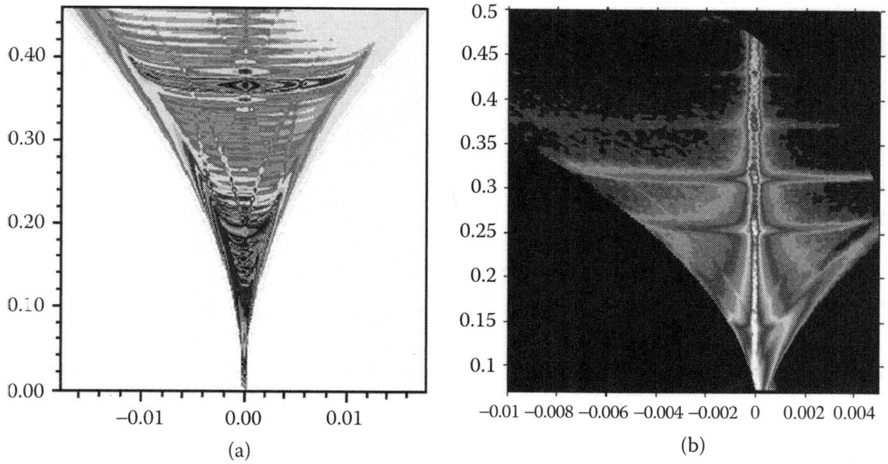

FIGURE 9.15
(a) Simulated reciprocal space map (RSM) for a Fe/Au multilayer in which the interface roughness is totally correlated. (Courtesy B.D. Fulthorpe.) (b) Experimental RSM for a sputtered Gd/Co multilayer.

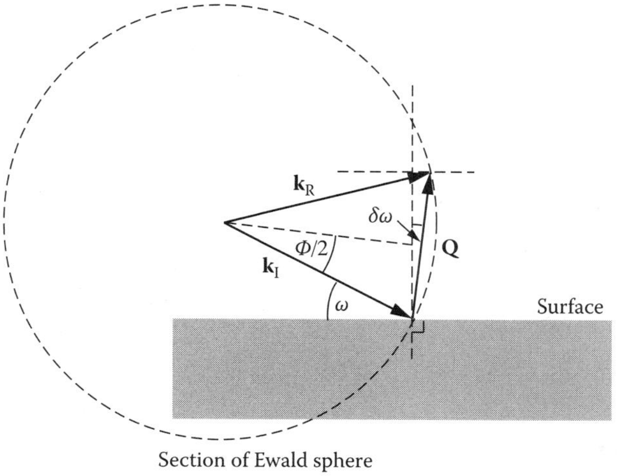

FIGURE 9.16
Scattering vector related to the angular displacement of the sample from the specular condition, as in a transverse scan. The Ewald sphere, of which we see a section, is the locus of all possible directions of the incident and scattered wave vectors.

9.6.5 Transformation from Angular Coordinates to Reciprocal Space Units

Conversion from angular coordinates (in which the scattered data are collected) to reciprocal space is straightforward. Consider, as in Figure 9.16, that the specimen is displaced an angle α' from the specular condition. Then it is easy to determine that

$$q_z = \left(\frac{4\pi}{\lambda}\right)\sin\left(\frac{\Phi}{2}\right)\cos(\delta\omega) \tag{9.33a}$$

and

$$q_x = \left(\frac{4\pi}{\lambda}\right)\sin\left(\frac{\Phi}{2}\right)\sin(\delta\omega) \tag{9.33b}$$

From these two equations, it can be seen why the Q_x range is small compared with Q_z. In Equation 9.33b there is the product of the sines of two small angles, while in Equation 9.33a there is only one.

9.7 Summary

- The theory of diffuse scattering forms a good description of the observed x-ray diffuse scatter from surfaces and interfaces.
- Diffuse scatter can:
  - Distinguish between topological roughness and composition grading
  - Determine structural parameters of the topology, which may affect film growth
  - Measure the conformality of multiple layers
- The theory is mathematically fairly complicated, and computational solutions are currently somewhat slow for in-line inspection.
- The analysis is very suitable for process development.

References

1. S.K. Sinha, E.B. Sirota, S. Garoff, and H.B. Stanley, *Phys. Rev. B*, 38 (1988) 2297.
2. M. Wormington, I. Pape, T.P.A. Hase, B.K. Tanner, and D.K. Bowen, *Phil. Mag. Lett.*, 74 (1996) 211–216.
3. L. Névot and P. Croce, *Rev. Phys. Applique*, 15, 761, 1980.
4. V. Holý and T. Baumbach, *Phys. Rev. B*, 49 (1994) 10668–10676.

10

Theory of XRD on Polycrystals

10.1 Introduction

The theory of x-ray scattering is highly practical. It is an accurate theory, based on a few sound assumptions. With implementations on personal computers, it may be used to interpret the atomic-scale structures of advanced industrial materials, and thereby to assist in process development and quality control. The x-ray metrologist who has a good grasp of the theory will therefore be able to design better measurements and to interpret them more accurately.

It is not our intention to provide full derivations of x-ray diffraction (XRD) theory, since this is primarily of interest to the specialist and may be found in many excellent books and reviews.[1–3] The x-ray metrologist needs a qualitative understanding of the scattering of x-rays by crystals in order to appreciate general features of experiment design and the interpretation of high-resolution rocking curves and images. He or she also needs specific numbers, such as the ideal rocking curve width or the penetration depth, for a particular specimen. Our aim is therefore simply to explain the aspects of the theory that are relevant to x-ray metrology and to summarize the important formulae.

It is conventional and useful to approach x-ray scattering theory on two levels, the so-called *kinematical* and *dynamical* theories. The simpler kinematical theory assumes that a negligible amount of energy is transferred to the diffracted beam, with the consequence that we can ignore rediffraction effects. This is fairly accurate for the geometry of diffraction in all cases, and is also satisfactorily accurate for the intensities when the scattering is very weak. Very thin crystals, surface scattering, and diffuse scattering are examples of weak scattering. When the scattering is strong, for example, for the diffracted intensities and rocking curve widths of near-perfect crystals, the kinematical theory breaks down. As the majority of the semiconductor materials used in industry behave as near-perfect crystals, it is easy to see why we must often use the dynamical theory for interpretation of data. Fortunately, many of the concepts of kinematical theory, such as structure factor and diffraction geometry, are also used in the dynamical theory.

10.1.1 Mathematical Health Warning

There is one major problem that we must point out here. It concerns the definitions of the wave vector **k** between Chapters 8 and 10. They are different. Grazing incidence scattering theory was developed by physicists who define $|\mathbf{k}| = 2\pi/\lambda$, whereas x-ray diffraction theory was derived by crystallographers who define $|\mathbf{k}| = 1/\lambda$. While in some ways it would be sensible to be consistent through this book, it would be flying in the face of two huge bodies of literature, each with their own conventions. In this chapter and Chapter 11 we will use the definition of crystallographers. Remember that the factor of 2π is then also missing from the scattering vector **Q**. The definition of **Q** remains the same. In order that there is no ambiguity, we will define certain parameters again, although these definitions may have been given in earlier chapters.

10.2 Kinematical Theory of X-ray Diffraction

The theory of x-ray reflectivity (XRR) described in Chapter 8 made the assumption that the material was continuous and could be described by a uniform electron density. Now we know that matter is composed of atoms and that the electron density is really very nonuniform. How is it that the theory of Chapter 8 works so well? The answer lies in the size of the scattering vector **Q**, which was defined in Equation 8.19 as the difference between the wave vector $\mathbf{k}_0$ of the incoming wave and the wave vector **k** of the outgoing wave, $\mathbf{Q} = \mathbf{k} - \mathbf{k}_0$. Then, a simple piece of geometry shows that $|\mathbf{Q}| = 2|\mathbf{k}| \sin\Phi/2$, where Φ is the scattering angle, that is, the angle between the exit and incident waves. A small scattering angle Φ means a small scattering vector **Q**. Recalling that $|\mathbf{k}| = 1/\lambda$, where λ is the x-ray wavelength, we see that a small **Q** corresponds to a large-length scale. For small scattering angle such as used in XRR, we are only sensitive to local variations in structure that are large compared with the x-ray wavelength. However, when the scattering angle becomes large, the corresponding length scale becomes comparable with the x-ray wavelength, and thus similar to the spacing of atoms in crystals; this is the regime of XRD.

As for the grazing incidence scattering treated in Chapters 8 and 9, we restrict the discussion to elastic scattering. We will make three key assumptions:

- The scattered intensity is very small. The loss of intensity due to rescattering is negligible, and thus the refractive index is unity.
- The point of observation is at a large distance compared with the dimensions of any coherently illuminated scattering volume.

- Scattered waves from different atoms are nearly parallel. We label these with the single wave vector, which from now on we call $\mathbf{k}_h$.

Conditions 2 and 3 are equivalent to the Fraunhofer or far-field approximations in ordinary optics. The coherently illuminated region with usual laboratory x-ray sources is a few microns across. We therefore expect this theory to be useful in the cases of weak scattering, but to be seriously awry for strong scattering.

The scattering from an atom, defined by the atomic scattering factor f_i, is usually defined relative to the scattering of an individual free electron. This is calculated as if the electron were a classical oscillator. It is set into forced oscillation by the radiation field of an incident x-ray and then reradiates in all directions at the same frequency as the incident wave frequency. This is termed elastic or Thompson scattering. For an x-ray beam, the intensity scattered by one electron relative to the incident intensity I_0 is the electron scattering factor, f:

$$f = \frac{I}{I_0} = \frac{Cr_e^2}{R^2} \tag{10.1}$$

where R is the distance of observation from the particle and r_e (the so-called classical electron radius or Thomson scattering length) is

$$r_e = \frac{e^2}{4\pi\varepsilon_0 mc^2} \tag{10.2}$$

where e is the electronic charge, m the rest mass of the electron, c the velocity of light, and ε_0 the permittivity of free space. (Its value is 2.82×10^{-15} m.) C is a factor dependent on the polarization. If the electric vector of the x-ray wave is perpendicular to the scattering plane, then $C = 1$; this is known as σ polarization. For π polarization, in which the electric vector is parallel to the dispersion plane, $C = \cos\Phi$. (Note that this is often described as $C = \cos 2\theta$ when the scattering is specular, θ being the incidence angle.) For an individual electron, the angular variation of the scattering arises only through this polarization term. It does not matter that electrons in a solid are not classical oscillators, since this is only a unit of reference, but it brings out the polarization behavior. In fact, electrons in an atom do behave surprisingly similarly to classical oscillators with respect to elastic x-ray scattering.

If all the electrons in an atom were concentrated at one point, then we should just multiply f by Z, the atomic number, to get the atomic scattering factor, f_i. This is a good approximation when the scattering angle Φ is small, as in Chapter 8, and thus all the scattering is nearly in-phase. However, atoms have finite sizes compared with x-ray wavelengths, and when $\Phi \neq 0$, we must add with regard to phase. The scattering in the direction defined

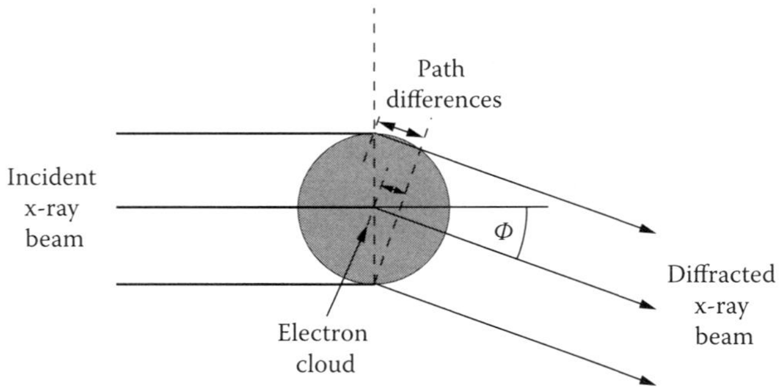

FIGURE 10.1
Scattering from distribution of electrons within an atom showing the optical path differences for various parts of the atom.

by **Q** from an electron at position vector $\mathbf{r}_j$ with respect to an arbitrarily defined origin within the atom is out of phase by a factor $exp(2\pi i\mathbf{Q}\cdot\mathbf{r}_j)$ with respect to an electron located at the origin (Figure 10.1). The total scattering factor f_i for an atom of type i containing Z_i electrons is then

$$f_i = \sum_{j=0}^{j=Z_i} f\exp(2\ i\mathbf{Q}\cdot\mathbf{r}_j) \tag{10.3}$$

Since the distribution of electrons in an atom is continuous, we express f_i as an integral rather than a summation. In units of f, or *electron units* (and hence itself dimensionless), this is

$$f_i = \int_{space} \rho(r)\exp(2\pi i\mathbf{Q}\cdot\mathbf{r})dV \tag{10.4}$$

The scattering vector is still given by $\mathbf{Q} = \mathbf{k}_h - \mathbf{k}_0$, where $\mathbf{k}_0$ is the incident beam vector, although we now label the scattered beam wave vector as $\mathbf{k}_h$ (Figure 10.2).

We see from Equation 10.4 that for very small angles of scatter $\mathbf{Q}\cdot\mathbf{r} \approx 0$ and hence

$$f_i(Q \rightarrow 0) = \int_V \rho(r)dV = Z_i \tag{10.5}$$

The atomic scattering factor for forward or very low angle scattering, neglecting correction for dispersion, is simply the atomic number Z_i.

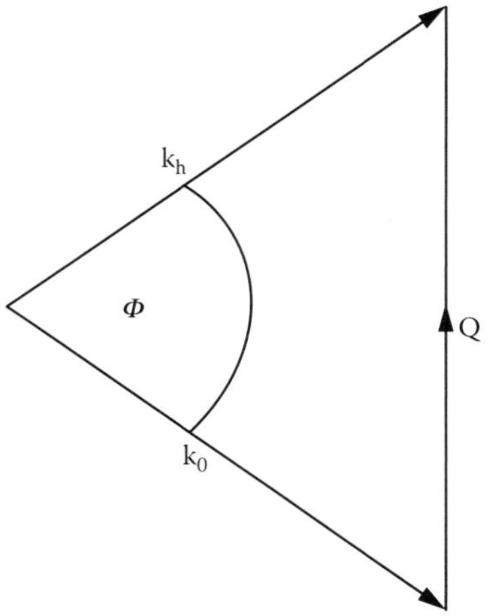

FIGURE 10.2
The vector relationship between the incident beam vector, $\mathbf{k}_0$, the scattered beam vector, $\mathbf{k}_h$, and the scattering vector, $\mathbf{Q}$. The directions of the vectors correspond to their directions in real space, and the magnitudes of $\mathbf{k}_0$ and $\mathbf{k}_h$ are both $1/\lambda$. Φ is the scattering angle.

It is seen from Figure 10.2 that the modulus of the scattering vector, Q, is given by

$$Q = \frac{2\sin(\Phi/2)}{\lambda} \tag{10.6}$$

In Equation 10.5, $\rho(r)dV$ is the probability that an electron lies in a volume element dV of the atom at a radial distance r from the nucleus. $\rho(r)$ is the electron probability density and is a meaningful and measurable quantity. The scattered amplitude is therefore the Fourier transform of the electron density. We see immediately that for a given atom the atomic scattering factor is a function only of $\mathbf{Q}$. Its angular dependence is important and is illustrated in Figure 10.3. The fall-off with scattering vector arises from the destructive interference of electrons spread across the atom. (The equivalent scattering factor for neutrons has no dependence on scattering vector, as the nucleus can be considered a point object.) Computation of the x-ray atomic scattering factors is not trivial, but fortunately they are tabulated for all elements in the *International Tables for X-Ray Crystallography*, published by the International Union of Crystallography.

10.2.1 Scattering from a Small Crystal

The difference between an amorphous and a crystalline solid is that the latter exhibits long-range order. The arrangement of atoms on such an ordered lattice differs from material to material, but fortunately, most materials of

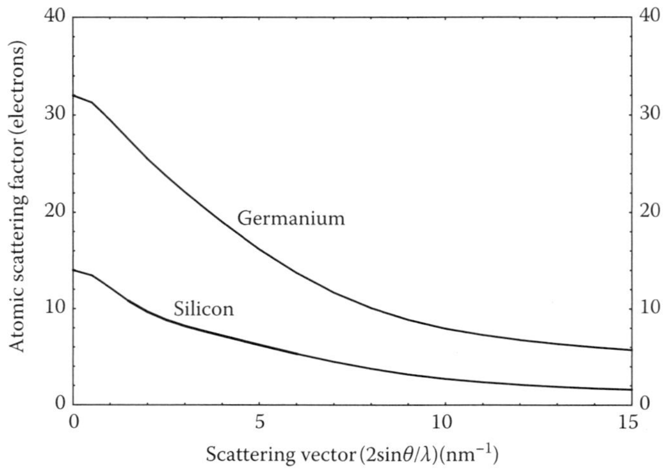

FIGURE 10.3
The variation of the atomic scattering factor, f_i, with scattering vector. Values for silicon (Z = 14) and germanium (Z = 32) are shown.

interest to the semiconductor industry have rather simple crystal structures and many have cubic structures.

For all crystal structures, we can define a unit cell that may be stacked together to form the crystal. Thus, just as we added the scattering from individual electrons to determine the atomic scattering factor, we now add the scattering from atoms coherently over a unit cell. Then we add the scattering from individual unit cells to obtain the scattering amplitude $A(Q)$ from the whole, macroscopic crystal. We can then write

$$A(q) = \sum_j \sum_i f_i(Q)\exp(2\pi i\,|\,\mathbf{Q}\cdot(\mathbf{T}_j + \mathbf{r}_i)\,|\,) \tag{10.7}$$

where $\mathbf{r}_i$ represents the position of atom i with respect to the origin of a unit cell and $\mathbf{T}_j$ is a real space vector defining the position of the jth unit cell. Because of the long-range order, the unit cells are located at integer multiples of the basis vectors of the structure. Equation 10.7 can be separated into two terms

$$A(Q) = [\sum_i f_i(Q)\exp(2\pi i\mathbf{Q}\cdot\mathbf{r}_i)]\sum_j \exp(2\pi i\mathbf{Q}\cdot\mathbf{T}_j) = F(Q)J(Q) \tag{10.8}$$

$F(Q)$ is called the structure factor for the unit cell, and we will come back to a discussion of its importance later. The function $J(Q)$ is known as the interference function. It can be evaluated by summing over all the unit cells within the crystal. We describe the crystal as a parallelepiped of side $n_1\mathbf{a}_1$,

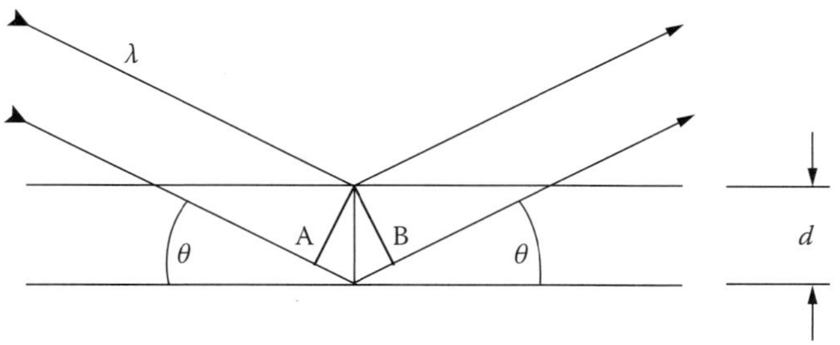

FIGURE 10.4
The Bragg condition for constructive interference.

$n_2\mathbf{a}_2$, and $n_3\mathbf{a}_3$, where $\mathbf{a}_1$, $\mathbf{a}_2$, and $\mathbf{a}_3$ are the unit vectors defining the unit cell and the n_i are the number of unit cells in each side of the parallelepiped. The interference function becomes

$$J = \sum_{n_1=1}^{n_1} \exp(2\pi\, i\, n_1\mathbf{a}_1 \cdot \mathbf{Q}) \sum_{n_2=1}^{n_2} \exp(2\pi\, i\, n_2\mathbf{a}_2 \cdot \mathbf{Q}) \sum_{n_3=1}^{n_3} \exp(2\pi\, i\, n_3\mathbf{a}_3 \cdot \mathbf{Q}) \qquad (10.9)$$

The exponential terms oscillate, and for most values of **Q**, we find that J is almost zero as a result of destructive interference. However, in the case of a crystalline solid, there are some very special values of **Q** for which the intensity adds in-phase. Lawrence Bragg realized that this corresponded to specular scatter from successive planes of atoms at a specific value of angle, now called the Bragg angle. We have already met this physical explanation in relation to the reflectivity from a periodic multilayer. The physics is identical; it is just that the spacing of the layers is now of the order of an angstrom, rather than a nanometer, and so the associated angle is now large.

Figure 10.4 shows the condition for constructive interference. The path difference between the waves scattered specularly from successive planes of atoms, spaced d_{hkl}, is $2d\sin\theta$. Constructive interference occurs when this in an integer number n of wavelengths λ. Thus,

$$n\lambda = 2d_{hkl}\sin\theta \qquad (10.10)$$

is Bragg's law, giving the condition for strong scattering. In terms of the scattering vector, Bragg's law is equivalent to the condition that $|\mathbf{Q}| = 1/d_{hkl}$. For appropriate beam and detector orientation, we can find the Bragg condition for different sets of atom planes. However, although pictorially useful and capable of telling us at what angles we can expect strong diffracted intensity, Bragg's law tells us nothing about the range over which that strong scattering occurs, nor of its intensity. To understand that, we must evaluate Equation 10.9, and to do this, we must introduce the reciprocal lattice.

10.2.2 The Reciprocal Lattice

We have already met reciprocal space in the context of grazing incidence diffuse scatter. Now we extend the ideas to take in crystal structures. We have deduced virtually all the rules already, but we formalize the description below.

- The dimension of reciprocal space is reciprocal length. We might choose a scale of, say, 10 mm represents 1 nm^{-1}.
- All directions in real space are preserved in reciprocal space.
- A reciprocal lattice vector is constructed for each *plane* of the Real-Space lattice as follows:
  a. The direction of the vector is perpendicular to the plane in real space.
  b. The magnitude of the vector is the inverse of the interplanar spacing in real space.
- The end of each such vector, starting from the origin, is a reciprocal lattice point.
- The reciprocal lattice is the set of reciprocal lattice points.

Thus, the reciprocal lattice axes are perpendicular to the (100), (010), and (001) planes in the real-space lattice. In cubic, tetragonal, and orthorhombic crystals it is also true that they are parallel to the [100], [010], and [001] directions, but this is not true in other crystal classes. The general formulae for the reciprocal space axes $\mathbf{a}_1^*$, $\mathbf{a}_2^*$, and $\mathbf{a}_3^*$ in terms of the real-space axes for a crystal whose unit cell is represented by the axes $\mathbf{a}_1$, $\mathbf{a}_2$, $\mathbf{a}_3$ are

$$\mathbf{a}_1^* = \frac{\mathbf{a}_2 \times \mathbf{a}_3}{\mathbf{a}_1.[\mathbf{a}_2 \times \mathbf{a}_3]}, \qquad \mathbf{a}_2^* = \frac{\mathbf{a}_3 \times \mathbf{a}_1}{\mathbf{a}_2.[\mathbf{a}_3 \times \mathbf{a}_1]} \quad \text{and} \quad \mathbf{a}_3^* = \frac{\mathbf{a}_1 \times \mathbf{a}_2}{\mathbf{a}_3.[\mathbf{a}_1 \times \mathbf{a}_2]} \tag{10.11}$$

We note that $\mathbf{a}_1^* \cdot \mathbf{a}_1 = 1, \mathbf{a}_2^* \cdot \mathbf{a}_2 = 1, \mathbf{a}_3^* \cdot \mathbf{a}_3 = 1$ and $\mathbf{a}_1^* \cdot \mathbf{a}_2 = 0, \mathbf{a}_2^* \cdot \mathbf{a}_3 = 0$, etc.

A plane in the real lattice can be described by its Miller indices *h*, *k*, and *l*, which correspond to the reciprocal of the distances from the origin that the plane cuts the axes $\mathbf{a}_1$, $\mathbf{a}_2$, and $\mathbf{a}_3$ (Figure 10.5).

The normal to the planes of Miller indices *h*, *k*, and *l* then corresponds in reciprocal space, through Equation 10.11, to a reciprocal space vector $\mathbf{h} = [h\mathbf{a}_1^*, k\mathbf{a}_2^*, l\mathbf{a}_3^*]$. The spacing of the planes in real space corresponds to the length of this vector in reciprocal space.

Reference to Figure 10.2 shows that the Bragg geometry and Bragg law are satisfied if the vector $\mathbf{Q} = \mathbf{h}$, where $|\mathbf{h}| = 1/d_{hkl}$, and the direction of $\mathbf{h}$ is perpendicular to the (*hkl*) planes.

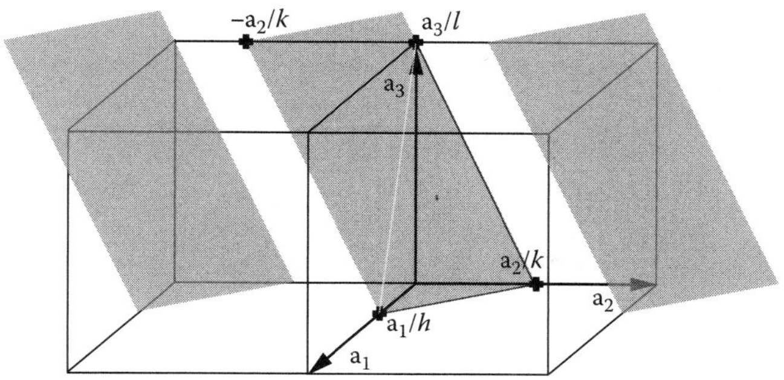

FIGURE 10.5
Intersection of the base axes $\mathbf{a}_1$, $\mathbf{a}_2$, and $\mathbf{a}_3$ by a plane of Miller indices h, k, and l. The actual plane shown is (321).

10.2.3 Intensity Diffracted from a Thin Crystal

We can express deviations from the Bragg law in terms of a deviation vector **s**, i.e.,

$$\mathbf{Q} = \mathbf{h} + \mathbf{s} \tag{10.12}$$

If **s** is small, in Equation 10.8, the *structure factor* $F(Q)$ associated with the set of atomic planes of Miller indices h, k, and l can be written as $F(h)$, or in slightly different notation,

$$F_{hkl} = \sum_i f_i \exp\{-2\pi i(hu + kv + lw)\} \tag{10.13}$$

where (uvw) are the fractional coordinates of the vector **r**, which runs from the origin of the unit cell to the atom of type i, whose atomic scattering factor is f_i, and the summation is over all atoms in the unit cell.

The structure factors of reflections in different cubic structures are given in Table 10.1. Note that the structure factors may be complex, but this depends upon the choice of origin of the unit cell. The intensity formulae always contain the modulus of the structure factor (or its equivalent in susceptibility, e.g., $\sqrt{\chi_0\chi_h}$) and also $1/V$, where V is the volume of the unit cell. The modulus formulae are given in this table. Also given are the actual moduli for aluminum, silicon, and gallium arsenide for CuKα_1 radiation, to show the effects of material and of the scattering angle; these have been divided by V to make direct comparison between the materials.

A number of the values are zero, for example, the 001 and 002 reflections in the diamond cubic structure possessed by silicon and germanium. These are often referred to as forbidden reflections. Reflections such as 002 and 222

TABLE 10.1

Values of $|F_{hkl}|$ for a Number of Useful Reflections in Cubic Structures, with Examples of $|F_{hkl}|/V$ in Electron Units per Cubic Å for One Crystal of Each Type

Reflection	fcc	Al	Diamond, Cubic	Si	Sphalerite	GaAs
001	0	0	0	0	0	0
002	$4f$	0.52	0	0	$4(f_1 - f_2)$	0.04
004	$4f$	0.36	$8f$	0.39	$4(f_1 + f_2)$	0.84
111	$4f$	0.55	$5.66f$	0.38	$4\sqrt{f_1^2 + f_2^2}$	0.82
222	$4f$	0.40	0	0	$4(f_1 - f_2)$	0.03
333	$4f$	0.28	$5.66f$	0.24	$4\sqrt{f_1^2 + f_2^2}$	0.5
011	0	0	0	0	0	0
022	$4f$	0.46	$8f$	0.45	$4(f_1 + f_2)$	0.99
044	$4f$	n.a.	$8f$	0.31	$4(f_1 + f_2)$	0.66
112	0	0	0	0	0	0
224	$4f$	0.30	$8f$	0.34	$4(f_1 + f_2)$	0.74
113	$4f$	0.40	$5.66f$	0.30	$4\sqrt{f_1^2 + f_2^2}$	0.66
115	$4f$	0.28	$5.66f$	0.24	$4\sqrt{f_1^2 + f_2^2}$	0.50

in sphalerite, the structure of a very important class of semiconductors such as gallium arsenide and indium phosphide, are not forbidden but are very weak, since they depend upon the difference between the atomic scattering factors of the constituent atoms. They are sometimes called quasi-forbidden and are very useful for emphasizing compositional differences in such materials, such as nonstoichiometry or superlattice structures. An example is the common GaAs/GaAlAs superlattice, in which the 002 reflection is most useful.

We can evaluate the interference function $J(Q)$ in Equation 10.9 by successively taking the terms, for example, that in $\mathbf{a}_1$, and writing in terms of $\mathbf{s}$. The interference function then simplifies drastically:

$$\mathbf{a}_1 \cdot \mathbf{Q} = \mathbf{a}_1 \cdot (\mathbf{h} + \mathbf{s}) = \mathbf{a}_1 \cdot \mathbf{h} + \mathbf{a}_1 \cdot \mathbf{s} \tag{10.14}$$

The first term, $\mathbf{a}_1 \cdot \mathbf{h}$, is the Miller index component h, which is an integer; thus, in the interference function it becomes unity since $exp(2\pi ni) = 1$. This corresponds to strong Bragg diffraction when $\mathbf{s} = 0$. The second term when put in the interference function becomes

$$J_1 = \sum_{n_1=1}^{n_1} \exp(2\pi\, i\, n_1 s_1) = \frac{\sin(\pi\, n_1 a_1 s_1)}{\sin(\pi a_1\, s_1)} \tag{10.15}$$

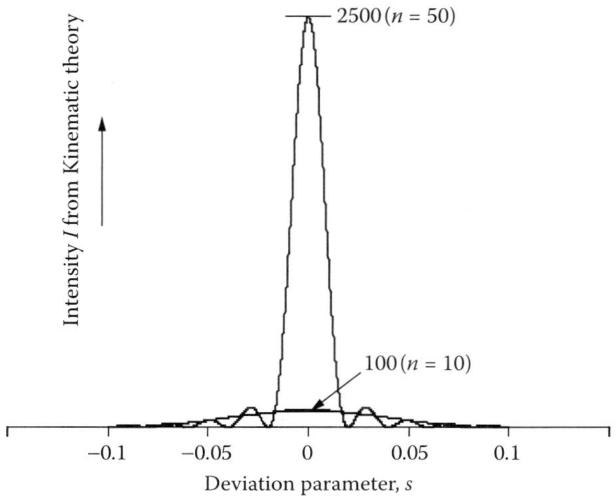

FIGURE 10.6
The relative scattered intensity as a function of one component of the deviation vector **s**, shown for $n_1 = 10$ and $n_1 = 50$. The peak sharpens rapidly with n_1. For $n_1 = 1000$, the peak value is 10^6 and the width only about 0.001 radian.

where s_i is the component of **s** along the first axis in the diagram of Figure 10.5. This is in fact the first axis of the reciprocal lattice, which in cubic crystals is parallel to the real-space axes of the lattice. The final expression for intensity is

$$I = F^2 J^2 = F_{hkl}^2 \frac{\sin^2(\pi n_1 a_1 s_1)}{\sin^2(\pi a_1 s_1)} \frac{\sin^2(\pi n_2 a_2 s_2)}{\sin^2(\pi a_2 s_2)} \frac{\sin^2(\pi n_3 a_3 s_3)}{\sin^2(\pi a_3 s_3)} \tag{10.16}$$

At small values of the deviation vector **s** the second part of the right-hand expression reduces to the product of three functions of $\text{sinc}^2(x)$ type. This is shown for one component in Figure 10.6.

Figure 10.6 brings out the following features of the scattering from a thin (or small) crystal:

- The scattering is only intense near **s** = 0, i.e., an exact Bragg reflection. The weak scattering from atoms is here reinforced by successive planes of atoms scattering in-phase. Strong intensity only occurs due to this reinforcement. Therefore, x-ray diffraction techniques integrate the scattering over many atomic layers.
- The first zero occurs at $a_i n_i s_i = 1$, i.e.; the width of the diffraction peak varies inversely as the number of atoms. We expect appreciable broadening of diffraction spots for crystallites less than a few tens of nanometers across, and this is indeed observed.

- The peak intensity, and also the integrated intensity, will be proportional to $|F_h|^2$.
- The scattered intensity is proportional to the volume of the crystal. This implies that the scattering from a thin epitaxial layer, large in area compared with the beam diameter, will be proportional to the layer thickness.

For large single crystals, extremely narrow rocking curves are predicted by the kinematical theory, and these are not found. Dynamical theory is required for these cases.

10.3 Determination of Strain

The angular position of a Bragg peak is determined by the spacing of the atom planes associated with a reciprocal lattice vector. If the material is macroscopically strained, there is a change in the Bragg planes associated with specific macroscopic directions. In principle, we can use the shift in the Bragg peak positions to measure the strain, and from it stress, in a material. This is of particular interest for polycrystalline thin films, where biaxial strain in the film plane is often present.

There is a huge literature associated with stress determination in thin films, and a comprehensive recent review has been given by Welzel and co-workers.[4] The classic method is to measure a diffraction pattern as a function of the angle between the diffraction vector and the crystal surface normal, ψ, and plot the lattice spacing of a given plane (deduced from the Bragg angle position) as a function of $\sin^2\psi$. The slope of the line gives the strain, and extrapolation to $\sin^2\psi = 0$ gives the stress-free lattice plane spacing. It is easy to see that this provides a measurement of the elastic strain components in the material, but the quantitative conversion to residual elastic stress can be quite complex. There are several reasons:

- The complexity of anisotropic elasticity in solids, which must be described by tensor equations.
- The difficulty of averaging properly the stresses and strains in a polycrystalline assembly. (Is it the stresses, the strains, or the displacements that are continuous across grain boundaries?)
- The complication of textured polycrystalline solids, which affects both the average distributions of lattice planes for measurement (expressed through the orientation distribution function) and the way in which the stress or strain averaging between grains is performed.

An example of biaxial stress determination in a thin film of Mo by the $\sin^2\psi$ method was given in Figure 1.9. In this case the beam penetrated the whole film and an average stress was found. However, techniques exist to measure not only the biaxial stress but its gradient, i.e., its value as a function of depth, and these were applied in the same study.[5] Figure 10.7 shows the geometry involved. The main method, known as the ϕ integral method, relies on the following key points:

- An incident beam at grazing or glancing incidence penetrates only a shallow layer of material. This penetration can be controlled by the angle of incidence, α. If there is a stress gradient toward the surface, then different strains will be measured at different values of α.
- The stresses and strains *must be continuous* in the sample. Therefore, they will be periodic in the rotation ϕ about the specimen normal, with a period of 2π.

Any periodic function may be expressed as a Fourier series, and in this case the Fourier components contain components of the strain tensor. Measurement of the $d(\phi, \psi)$ values for at least two (preferably more) values of ψ is sufficient to determine all six independent components of the strain tensor.[6,7] With addition of α variation, the complete triaxial stress can be determined as a function of depth. In practice, the stress state is often biaxial (absence of shear stresses can be recognized by the absence of ψ splitting of lattice spacings between positive and negative values) and the mathematics then simplifies.

Figure 10.8 shows the data and final stress gradient determination for the case of the Mo thin film previously discussed. It is seen that stress gradients can be determined in thin films with a depth resolution of around 10 nm.

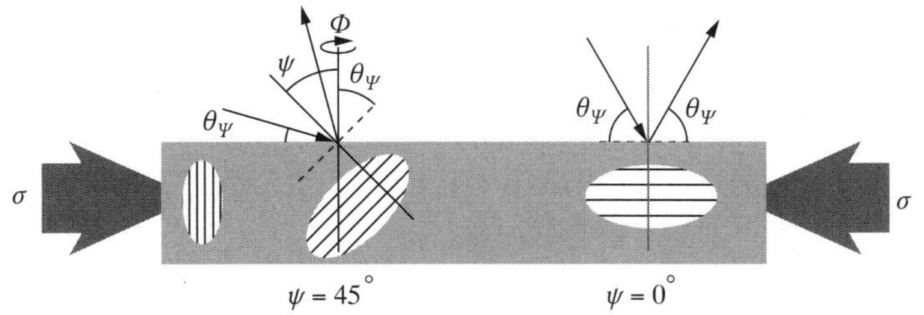

FIGURE 10.7
Definition of the geometry used in measurement of strain by XRD. (From Welzel, U. et al., *J. Appl. Cryst.*, 38, 1–29, 2005.)

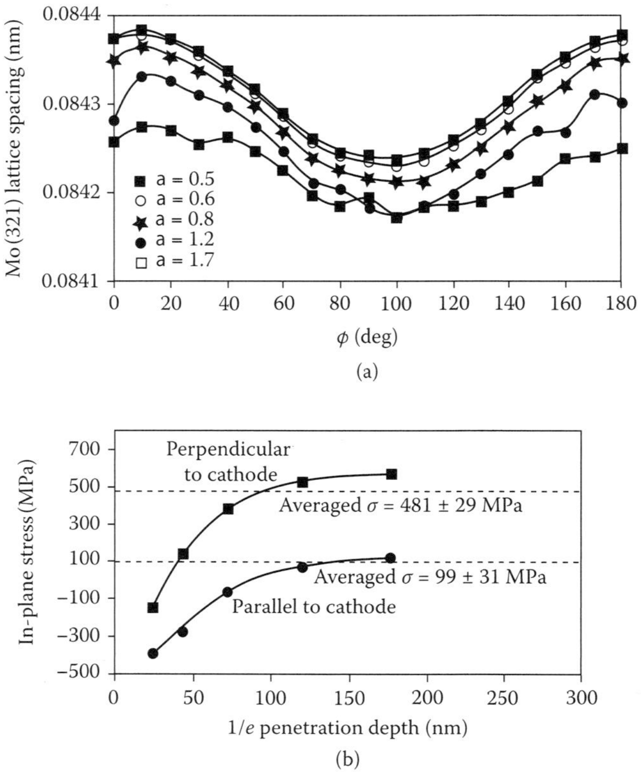

FIGURE 10.8
An example of measurement of stress gradient in Mo by the φ integral method. (a) The Mo (321) lattice spacing plotted as a function of the grazing incidence angle, α, and of the rotation about the specimen normal, φ. (b) The in-plane stresses calculated as a function of the $1/e$ penetration depth. The average values obtained by the $\sin^2\psi$ method are also shown. (From Ballard, B.L. et al., *Adv. X-Ray Anal.*, 37, 189–196, 1994. With kind permission from Springer Science and Business Media.)

10.4 Determination of Grain Size

The strain dispersion is associated with the dislocation density, a product of plastic deformation that has occurred. This strain dispersion results in a broadening of the Bragg peaks, and the magnitude can be determined by differentiation of Bragg's law (Equation 10.10). The resulting distribution of Bragg angles, $\Delta(\theta)$, is given by

$$\Delta(\theta) = \tan\theta\,(\Delta d/d) = \tan\theta\,\Delta\varepsilon \qquad (10.17)$$

where Δd is the dispersion in the interplanar spacing d. The fractional dispersion $\Delta d/d$ is the strain dispersion $\Delta\varepsilon$. In terms of the measured detector angle 2θ, this becomes

$$\Delta(2\theta) = 2\tan\theta \, \Delta\varepsilon \tag{10.18}$$

Note, as with measurement of the elastic strain, the measurement is the deformation in a direction normal to the crystal planes probed, i.e., in the direction of the associated reciprocal lattice vector.

Broadening of the Bragg peaks in a powder diffraction pattern also arises from a small grain size, the peak widths of nanocrystalline materials being typically up to a degree or so. As we see from Section 10.2.3, the peak width scales with the number of atoms, i.e., the grain diameter, in the direction normal to the diffracting planes. Note that where the grain size is nonspherical, the direction of measurement with respect to the material is important. An example is of a thin-film polycrystalline material, where film thickness, not the crystallite size, may determine the peak width if measurements are made in the symmetric geometry "$\theta - 2\theta$" scan, where the diffraction vector is always perpendicular to the film surface.

From Figure 10.2 we have

$$Q/2 = k \sin\theta \tag{10.19}$$

and thus the small deviation s is given in terms of the angular deviation $\Delta\theta$ by

$$s/2 = k \cos\theta \, \Delta\theta \tag{10.20}$$

The zero intensity in the interference function occurs when $s_i = (n_i a_i)^{-1}$, and as the grain size $L = n_i a_i$, the zero in intensity occurs at $s = 1/L$. For a $sinc^2(x)$ peak shape, the value at which the first zero occurs is approximately the full-width-at-half-height maximum (FWHM). Thus, in terms of the FWHM in 2θ, Equation 10.20 becomes

$$\Delta(2\theta) = \frac{\lambda}{L\cos\theta} \tag{10.21}$$

From Equations 10.18 and 10.21 we obtain the Williamson–Hall equation that we presented in Chapter 4 (Equation 4.2):

$$\Delta(2\theta)\cos\theta = 2\Delta\varepsilon\sin\theta + \frac{\lambda}{L} \tag{10.22}$$

We see then that there are two effects that broaden the Bragg peaks: grain size and strain dispersion. However, the dependence of each on Bragg angle differs, and from a Williamson–Hall plot for a number of Bragg reflections, the two effects can be distinguished. The grain size, or more accurately the diffracting domain size, is determined by the intercept, while the slope determines the strain dispersion.

10.5 Texture

For a randomly dispersed powder, the relative integrated intensities under the various Bragg peaks can be determined from the square of the structure factor and the multiplicity factor. The latter factor arises from their being more than one plane of the type {*hkl*}. In diffraction from polycrystalline materials, strong scattering occurs when one or more crystallites happen to be oriented in such a way as to satisfy the Bragg condition, not at a specific position of the sample. For the {111} family of reflections, we have $(111),(\bar{1}11),(1\bar{1}1),(11\bar{1})$ as possible planes, and hence multiplying by 4 the probability that such a chance orientation occurs.

This approach works for randomly oriented powders, particularly when the sample is spun to improve the averaging of grain orientations, but it can fail seriously for thin-film polycrystalline materials. Materials such as Ta grow on silicon as columnar crystallites, the axis of the column (normal to the substrate plane) being predominantly [111]. Although the orientation of the crystallites within the plane of the substrate is random, the strong [111] texture makes determination of composition by analysis of the intensity under various Bragg peaks very unreliable. One the other hand, the degree of texture in a thin film can be assessed by the relative deviation from the expected peak heights of materials of known composition.

As an example, let us consider the [111] textured film. In a symmetric $\omega-2\theta$ scan, for example, in the Bragg–Brentano geometry, the diffraction vector is always normal to the film plane. When the detector is at the 2θ value to satisfy the Bragg equation for a particular *hkl* reflection, if there are crystallites with the (*hkl*) plane oriented parallel to the film, then a Bragg peak is observed. If there are no crystallites so oriented, there will be no Bragg peak, despite being at the correction detector orientation. In the case of the 111 reflection, there will be a very strong Bragg peak for the [111] textured film. However, if we look for the 220 reflection, we will find it very weak, if not absent. There are very few crystallites oriented so as to make the diffraction condition.

On the other hand, if we use grazing incidence in-plane x-ray diffraction (GIIXD), where the diffraction vector is almost in the plane of the sample, we will find a very strong 220 reflection, as there are many crystallites in a [111] textured film with a {110} plane normal to the film surface (for example, $[1\bar{1}0].[111]=0$). Conversely, the 111 reflection will be absent or at least very weak. It is important to note that the broadening of the Bragg peaks in a polycrystalline GIIXD pattern relates to the grain size and strain dispersion in the film plane. A Williamson–Hall plot from GIIXD data enables these two in-plane components to be distinguished. For thin films, symmetric geometry $\omega - 2\theta$, e.g., Bragg–Brentano, measurements *do not* provide information on in-plane grain size or strain.

Texture measurements in thin films can be carried out by fixing the detector at a constant angle (to pick up the intensity of diffraction of a specific plane) and scanning the sample over as large a solid angle as possible. The best scan for this is a loop scan, with tilt χ as the outer loop and rotation ϕ as the inner loop (see the inset on p. 8 for the notation), since this covers the whole of the pole figure. More limited scans can be done to measure restricted regions. There are a number of ways of plotting the orientation distributions, but they all show the relative distribution of the chosen Bragg planes within the sample. The pole figure (either as a contour map or as a three-dimensional view) is the most common; an excellent example was shown in Figure 1.10. To obtain the complete description of the (average) grain orientations, the orientation distribution function (ODF) is used. This plots the density of particular planes in orientation space and requires three independent pole figures (i.e., with different reflections) for its calculation.

10.6 Reciprocal Space Geometry

We introduced reciprocal space in the previous section. It is so very helpful in the interpretation of many diffraction experiments that we need to understand it more fully. We shall use it extensively in the discussion of triple-axis experiments, in which reciprocal space mapping is an essential technique. Figure 10.2 may be slightly extended to show a most useful construction, that of the Ewald sphere. Since we have elastic scattering, it is always true that the magnitudes of $\mathbf{k}_0$ and $\mathbf{k}_\mathrm{h}$ are both $1/\lambda$. A sphere of radius $1/\lambda$ can therefore define all possible incident and scattered beam vectors. The incident beam vector runs from the center of this Ewald sphere to the origin, the scattered beam vector runs from the center to any point on the surface of the sphere, and the scattering vector runs from the origin to the end of the scattered beam vector. Figure 10.9 shows a two-dimensional section of this three-dimensional construction. In this figure the scattering vector $\mathbf{Q}$ has been made to coincide with a vector $\mathbf{h}$ that satisfies the Bragg law (Equation 10.12), and we expect strong diffraction.

Of course, many reflections are possible from a regular lattice, and if the set of vectors such as $\mathbf{h}$ is represented, then we can graphically visualize all the reflecting planes and consequent reflected directions in the crystal. Some examples will illustrate the concept. These are all drawn to scale for the silicon lattice and structure. In Figure 10.10 a high-resolution experiment is shown.

It is quite convenient to use reciprocal space to show the accessible reflections in a given experiment. These are shown in Figure 10.11, for a silicon

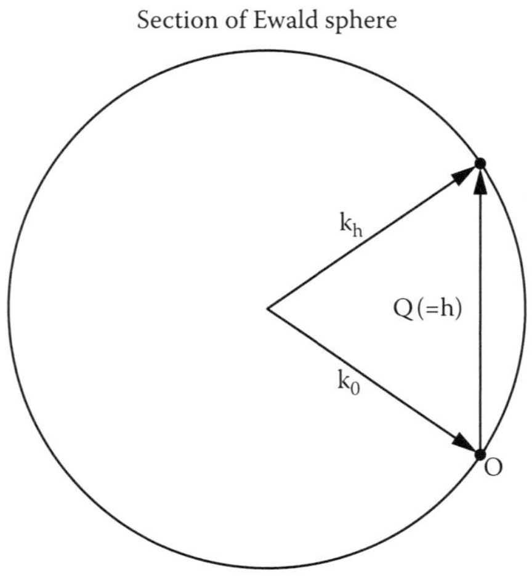

FIGURE 10.9
The Ewald sphere construction in reciprocal space.

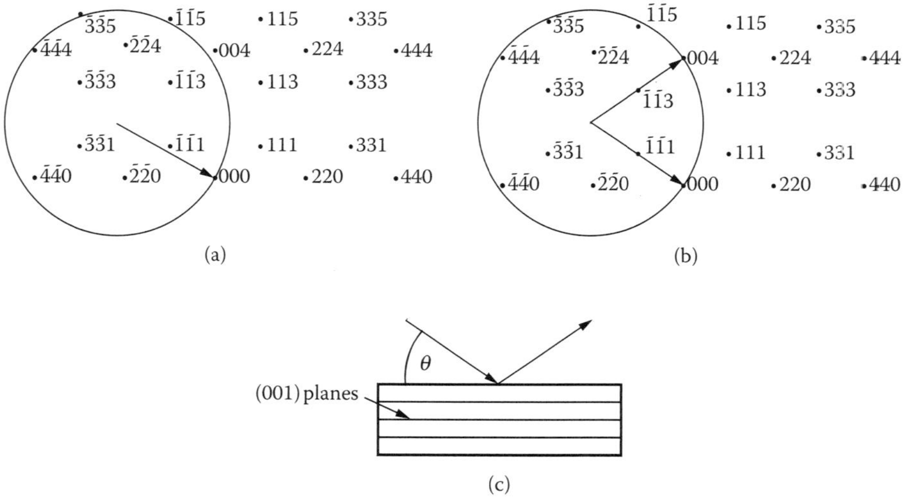

FIGURE 10.10
The reciprocal space representation of a high-resolution experiment. In (a) the crystal is not yet aligned to the Bragg position and no diffracted beam occurs. In (b) the incident beam has been rotated so that the Ewald sphere falls on the 004 reciprocal lattice point, and a diffracted beam ensues. The Ewald sphere is to scale for CuKα, and the reciprocal lattice is to scale for silicon.

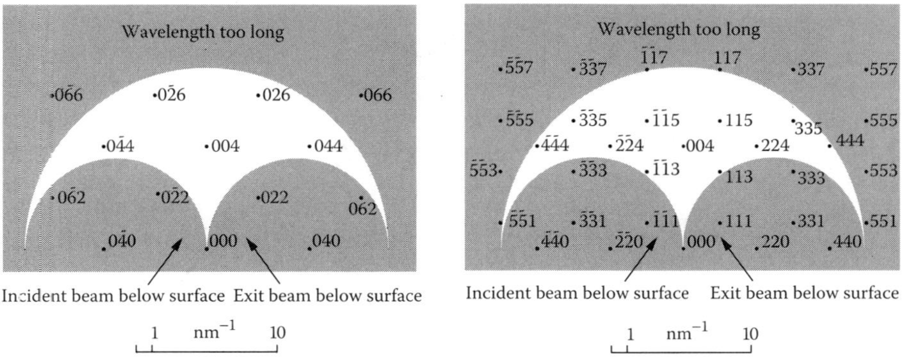

FIGURE 10.11
The accessible, allowed reflections in a high-resolution experiment, shown in reciprocal space, for a silicon specimen with a (001) surface, using CuKα radiation. (a) Incident beam in (100) plane, reciprocal lattice section perpendicular to [100]. (b) Incident beam in (110) plane, reciprocal lattice section perpendicular to [$1\bar{1}0$]. Case (b) would occur if the incident beam were parallel or perpendicular to the plane of the (110) flat or the direction of the notch cut on semiconductor wafers. Many more reflections are available in case (b). The semicircular segments showing accessible and inaccessible reflections are sections of hemispheres whose axes are vertical on this diagram.

specimen, with a (001) surface plane, and CuKα radiation, for two of the orientations of the incident beam. The large semicircle contains all the points that are cut by the Ewald sphere as the incident beam is rotated 180° from just grazing the surface in one direction to just grazing the surface in the opposite direction. The small semicircle on the left contains points that cannot be accessed in reflection because the incident beam would enter from below the crystal surface. The small semicircle on the right likewise cannot be accessed in reflection as the diffracted beam exits through the crystal. These two regions are accessible in transmission. The reflection and transmission conditions are often called *Bragg case* and *Laue case*, respectively.

In the powder or polycrystalline thin-film diffraction experiment, a fixed wavelength and incident beam direction are used, and the random orientation of the powder grains means that the reciprocal lattice is rotated about all angles, centered on the origin. A diffracted beam ensues whenever a reciprocal lattice point cuts the Ewald sphere. This is shown in two dimensions in Figure 10.10. This construction also makes it easy to see when reflections overlap, for example.

These examples show the great practical utility of the Ewald sphere construction. At a summer school in 1974, we once heard Paul Ewald say, some 60 years after he laid the basis of x-ray scattering theory, that he wished people had named something else after him, as he felt that it was such a trivial idea.

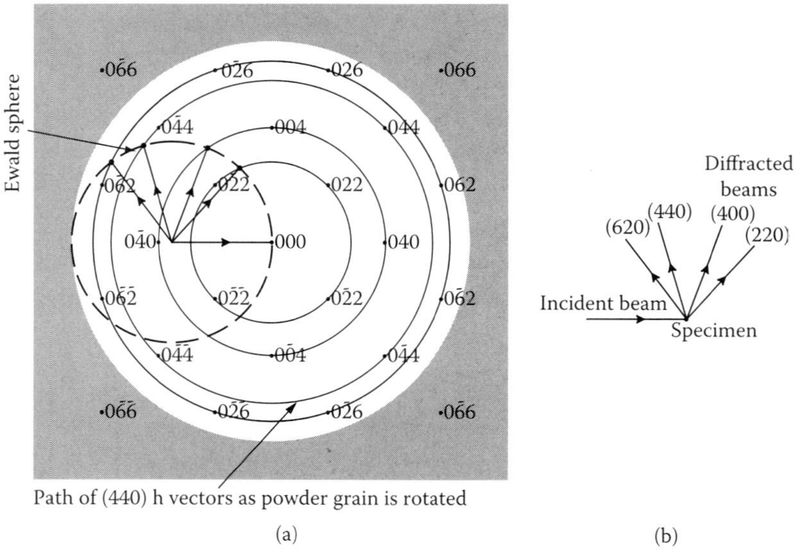

FIGURE 10.12
The powder or polycrystalline thin-film diffraction experiment. (a) Reciprocal space notation. The Ewald sphere is fixed, and the lattice is rotated about all angles about the origin. Only the rotations about [100] are shown in this two-dimensional section. Intersections with the Ewald sphere define the diffracting conditions. (b) The corresponding diffracted beams in real space.

10.7 Summary

- Strong scattering from polycrystalline materials occurs at angles where the Bragg condition is satisfied.
- For thin films and small crystallites, kinematic theory is a sufficient description of the diffracted positions and intensities.
- The Bragg peaks are not nearly as sharp, nor as intense, as predicted by kinematical theory for thicker crystals.
- The width of the peaks depends on the grain size and the strain dispersion within the grains.
- The effects of strain dispersion and grain size can be separated by analysis of different reflections.
- The positions of the peaks can be used to determine internal stresses in polycrystalline materials.
- Texture affects the relative intensities of diffraction lines from thin films.
- The Ewald sphere is an elegant means of identifying whether specific Bragg reflections are accessible from particular specimens.

References

1. B.E. Warren, *X-Ray Crystallography*, Addison-Wesley, Reading, MA, 1969.
2. A. Authier, *Dynamical Theory of X-Ray Diffraction*, Oxford University Press, Oxford, 2001.
3. U. Pietsch, V. HolΔ, and T. Baumbach, *High-Resolution X-Ray Scattering*, 2nd ed., Springer, New York, 2004.
4. U. Welzel, J. Ligot, P. Lamparter, A.C. Vermeulen, and E.J. Mittemeijer, *J. Appl. Cryst.*, 38 (2005) 1–29.
5. B.L. Ballard, P.K. Predecki, and D.N. Braski, *Adv. X-Ray Anal.*, 37 (1994) 189–196.
6. W. Lode and A. Peiter, *Metall.*, 35 (1981) 758–762.
7. C.N.J. Wagner, M.S. Boldrick, and V. Perez-Mendez, *Adv. X-Ray Anal.*, 26 (1983) 275–282.

11

High-Resolution XRD on Single Crystals

11.1 Introduction

The greatest use of high-resolution diffractometry in industry is the characterization and metrology of epitaxial structures on semiconductors. As explained in Part 1, the main need is to measure thickness and strain in the layers. From strain, the composition may be deduced for a binary or ternary alloy, and it is required for strained silicon layers.

Such epilayers may also be mismatched, misoriented, defective, nonuniform, and bent. These defects affect device performance and production yield. Residual strains in the layer can be correlated with poor device performance or with degradation in service. Dislocations in the interface, from relaxation, are particularly damaging. On the other hand, a controlled relaxation may be required in order to make, for example, a substrate for a thin strained silicon layer. In the case of III-V materials, the layers are nowhere near as perfect as silicon, either in defect concentration or in composition uniformity. The defects affect carrier lifetimes and may act as nonradiative recombination centers. Composition variation affects the band gap as well as the mismatch. It is therefore essential in the development phase of a device to determine whether the defects generated in a particular production process will permit adequate yield of good-quality devices. Later on, in manufacture, it is necessary to determine whether the process is under adequate quality control by a selective, rapid test that verifies the key parameters.

In this chapter we shall first provide a summary of the relevant theory. The emphasis is on the theoretical basis for the measurements described in Part 1 of this book, and on their reliability. We refer the reader to other books for derivation of the theory itself.

We shall then see how the basic parameters can be obtained and interpreted from x-ray rocking curves. In a fab line, the interpretation is always performed by modeling and simulation of diffraction curves, but it is useful to know the basic effects of various parameters in order to design effective recipes for modeling.

During the development of a device or process, it is often necessary to characterize the layer structure in more detail than is required for quality control. The reciprocal space map (RSM) is most useful for this stage, and the following section is devoted to the use and interpretation of RSMs.

Finally, some of the structures now being made, for example, in strained silicon, depend upon thin strained surface layers. These are sometimes isolated, as in silicon on insulator (SOI) processes, and sometimes grown on SiGe or other layers. In order to separate the signal from such layers from those of the substrate or of other layers, it may be necessary to confine the x-ray beam to the surface layer itself. This is achieved by grazing incidence in-plane x-ray diffraction (GIIXD), which is discussed in the final part of this chapter.

11.2 Dynamical Theory of X-ray Diffraction

In Chapter 10, we introduced the kinematical theory of x-ray diffraction. This works well for very thin films and materials or for containing small crystallites, but it is unsatisfactory for predicting intensities of thicker epitaxial films or single crystals. We can see immediately that there is a problem for large crystals. The predicted peak width scales inversely with the grain size (or film thickness). But such a reduction in Bragg peak width is not observed, and it is not just a matter of instrument resolution.

The assumptions made in deriving the kinematical theory are equivalent to requiring that no energy is transferred into the diffracted beam. There is necessarily saturation in the diffracted intensity as the crystal thickness increases, and many additional interference effects caused by the presence of strong wave fields that are not predicted by kinematical theory. However, thanks to theoretical studies over the past 90 years, the dynamical theory, which corrects these gross errors, is very well understood. Moreover, the recent arrival of packages providing numerical solutions on accessible personal computers means that industrial scientists can obtain precise, quantitative descriptions of the scattering that can be used in process development and quality control.

There are two approaches to calculation of the diffracted intensity from a thick, perfect crystal. One is to solve Maxwell's equations for a periodically modulated medium. Solutions are found to the wave equations that correspond to the modes that propagate unchanged inside the crystal. These normal modes are referred to as wave fields and are not plane waves, but Bloch waves. This classical treatment of x-ray dynamical diffraction theory gives a detailed understanding of the energy flow within the crystal and has been shown to be in excellent agreement with experiments. Since the theory is well treated in many excellent texts[1–11] we will not discuss it in this book. Its main difficulty lies in the requirement to match wave fields at every

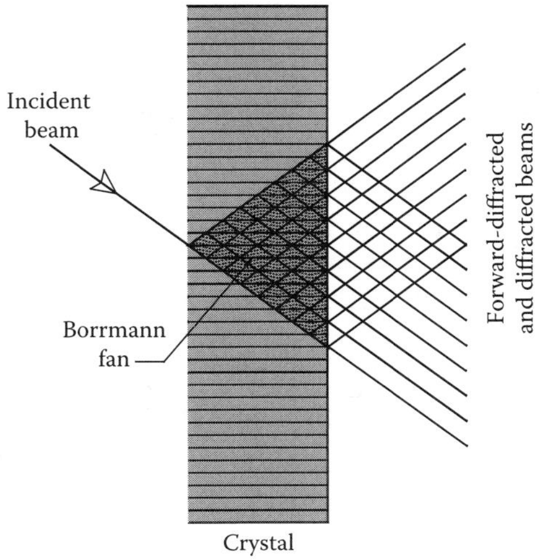

FIGURE 11.1
The diffraction and rediffraction of an x-ray beam from a set of reflecting planes. The triangle bounded by the incident beam and the diffracted beam from the entry surface is called the Borrmann fan.

boundary. Despite the use of matrix methods, this becomes cumbersome for multilayer structures.

The second approach is a multiple-scattering theory in which the waves inside the crystal are assumed to be plane waves, but changing in nature as they pass through the crystal. In essence, the theory is based upon the idea that the diffracted beam from a set of reflecting planes is at the correct angle to be rediffracted by the same planes. This is illustrated in Figure 11.1. It is seen that the energy is spread throughout a triangular region in this section, known as the *Borrmann fan*. We expect this to result in a complicated expression for intensity, as indeed it does, though it can easily be implemented on a personal computer if the region is uniform. If it is nonuniform, as are many interesting industrial materials, this coupling of the diffracted and *forward-diffracted* beams (we must no longer think of the incident beam as transmitting unchanged through the crystal) must be treated locally around the inhomogeneities. The *Takagi–Taupin* theory is based upon the formulation and solution of a coupled pair of differential equations that represent the changes in amplitude in each of the forward-diffracted and diffracted directions.

The Takagi–Taupin theory and the simulation methods that it allows are extremely powerful and useful. However, in this approach we lose everything but the numerical amplitudes and intensities. We do not know how beams propagate through the crystal, how and when they are likely to interfere, or what parameters control rocking curve width and shape; in other words, the physics is buried.

Ideally, we want a theory that would take in our experimental rocking curves and give out the structure. That is not possible; it is one of the class of inverse problems with limited experimental information (in particular, the phase of the x-rays and the limited sampling of reciprocal space that a rocking curve provides). What we do have is a practical theory that can be implemented on personal computers for simulating the rocking curve of a material whose structure is known. Comparison of the features and then the intensities of the simulated and experimental curves permits iterative refinement of the simulated structure. Simulation of rocking curves is an extremely powerful method of interpretation of complex structures. In addition, it is very valuable for the design of experiments, optimization of data collection strategy, and education of new researchers and operators in high-resolution diffractometry.

11.2.1 The Takagi–Taupin Generalized Diffraction Theory

This generalized diffraction theory, developed independently by Takagi[12,13] and Taupin,[14] can be used to describe the passage of x-rays through a crystal with *any type* of lattice distortion. As we emphasized in the previous section, the theory is not formulated in terms of wave fields, but is a multiple-scattering theory. Its basis is the same as the dynamical theory of electron diffraction, and the famous Howie–Whelan equations are an approximation of the Takagi–Taupin equations. The theory assumes that x-rays are propagating as plane waves and that scattering is occurring both into and out of the diffracted beam. Mathematically, we can always keep the account straight by adding the correct phase factors into the scattering from forward to diffracted wave and vice versa.

The important thing about this theory for simulating the rocking curves of multiple layers and multilayers superlattices is that we assume a single wave vector, and no matching at the boundaries needs to be done. So at the expense of understanding what physically goes on inside the crystal, we have a mathematical calculating method that works splendidly. The theory can be applied equally well to deformed and distorted crystals as to a perfect crystal. It has therefore become the most powerful method both for interpreting rocking curves of complex epilayer systems and for simulating the contrast of defects such as dislocations in x-ray topographs. Using simulation techniques, the scattered x-ray intensity can be related to the microscopic lattice strains in the crystal.

The wave amplitude inside the crystal is described in a differential form; D_o and D_h are the total amplitudes of the wave in the forward and diffracted beam directions, respectively, which may be slowly varying functions of position. How slowly? Surprisingly, the theory works very well for changes as abrupt as stacking faults and relaxed epilayer interfaces. $\mathbf{K}_o$ and $\mathbf{K}_h$ are the incident and scattered wave vectors, respectively, inside the crystal. We take $|\mathbf{K}_o| = nk$, where n is the refractive index far from Bragg reflection.

As is usual and accurate in most x-ray diffraction, we use the two-beam approximation; that is, only the forward and diffracted beam wave fields have appreciable intensity. With these assumptions, Takagi and Taupin took a modified Bloch wave representation of the wave field and obtained two coupled second-order partial differential equations expressed along the forward and diffracted beam directions $\mathbf{s}_o$ and $\mathbf{s}_h$ (these are unit vectors in the directions of $\mathbf{K}_o$ and $\mathbf{K}_h$):

$$\frac{\lambda}{i\pi}\frac{\partial D_0}{\partial \mathbf{s}_0} = \chi_0 D_0 + C\chi_{-h} D_h \tag{11.1}$$

$$\frac{\lambda}{i\pi}\frac{\partial D_h}{\partial \mathbf{s}_h} = (\chi_0 - \alpha_h) D_h + C\chi_h D_0 \tag{11.2}$$

where C is the polarization factor and α_h represents the deviation of the incident wave from the exact Bragg condition. This is a key parameter, as it is this that we vary when we scan a specimen to collect the rocking curve. We treat mismatched layers, graded layers, and all kinds of defects by their effect on a local distorted reciprocal lattice vector, and hence on the deviation parameter. Physically, the Takagi–Taupin equations tell us that the rate of change of the forward beam with distance is proportional to the amplitudes of the forward and scattered waves. Similarly, the rate of change of the scattered wave also depends on the amplitudes of the forward and scattered waves. Unsurprisingly, the Fourier components of the susceptibility χ_0 and χ_h, which determine the strength of this interchange, are proportional to the structure factors in the forward and scattered directions,

$$\chi_h = -\frac{r_e \lambda^2}{\pi V} F_h \tag{11.3}$$

where V is the volume of the unit cell.

For a perfect, uniform crystal, whether in bulk or as a thin layer, the Takagi–Taupin equations can be solved exactly, as given in the next section. For the general case with multiple layers, however, it is necessary to integrate them numerically. The concepts of the dispersion surface are lost, and we cannot tell directly in which directions wave fields are propagating. They do give *directly* the intensities of the direct and diffracted beams emerging from the crystal, and all interference features are preserved.

All that we need to do now is to solve the Takagi–Taupin equations and plug in the deviation parameter to predict any intensity. For the general distorted crystal, numerical solution is used. The most common is that over a grid of points, which may be distorted in any way, for example, to model the strains caused by a dislocation or precipitate at a given depth, and hence to simulate the x-ray topographic image. There have been many attempts to solve Takagi's equations analytically. However, the only solutions of real interest to us here are for the perfect crystal and the thin layer.

11.2.2 Thin-Layer and Substrate Solutions

Halliwell et al.[15] obtained an important solution of the Takagi–Taupin equations for a uniform layer of known composition, structure, and thickness. This allows any one-dimensional strain distribution to be obtained by splitting up the crystal into lamellae of constant strain. The solution is expressed in terms of the variables

$$A = C\chi_{-h} \tag{11.4}$$

$$B = \frac{(1-b)\chi_0}{2} + \frac{\alpha_h \pi}{2} \tag{11.5}$$

$$D = \frac{\pi}{\lambda \gamma_0} \tag{11.6}$$

$$E = -Cb\gamma_h \tag{11.7}$$

where $b = \gamma_o/\gamma_h$. We also write

$$F = \sqrt{BB - EA} \tag{11.8}$$

These are all complex variables since the susceptibilities are complex. We obtain the amplitude ratio at the top (exit) of the layer, X, in terms of that at the bottom (entrance), X':

$$X = \frac{X'F + i(BX' + E)\tan(DF(z - w))}{F - i(AX' + B)\tan(DF(z - w))} \tag{11.9}$$

The variable z is the depth above the depth w, at which the amplitude ratio is the known value X', in effect the thickness of the layer or lamella. We know the amplitude ratio deep inside the crystal; here, both the diffracted intensity and the incident amplitude are zero, but since their ratio must always be <1, then the amplitude ratio $X' = D_h/D_o(z)$ must also be zero. Using the above we may derive the reflectivity of a thick crystal:

$$X = -\left(\frac{B + F\,sign\left(\mathrm{Im}(F)\right)}{A}\right) \qquad (11.10)$$

The parameters A, B, D, and E are of fundamental importance. They depend upon the crystal susceptibilities χ_o, χ_h, and χ_{-h}, the cosines of the inclination angles of the incident and diffracted beams with the inward-going surface normals (γ_o and γ_h, with the asymmetry factor $b = \gamma_o/\gamma_h$), the polarization factor C, the wavelength λ, and the deviation parameter α_h. Except for α_h, we may look up or calculate all of these from the composition and structure of the material and the geometry of the experiment. Thus, the problem is reduced to the calculation of the deviation parameter.

11.2.3 Calculation of Strains and Mismatches

At reasonably large angles the deviation parameter is given by:[16]

$$\alpha_h = -2\Delta\theta_h \sin 2\theta_B \qquad (11.11)$$

where $\Delta\theta_h$ is the local deviation from the exact Bragg angle, taking account of lattice strains (with the sign convention that incident angles below the Bragg angle are negative), and θ_B is the local exact vacuum Bragg angle. At small angles, which are very important for analyzing thin layers, a better expression is:[17]

$$\alpha_h = \left[\gamma_0\left(\gamma_0 - \gamma_h - 2\sin\theta_B\cos\varphi\right) + \tfrac{1}{2}\chi_0\left(1-b\right)\right] \times \left[|b|^{1/2}\, C\left(\chi_h\chi_{\bar{h}}\right)^{1/2}\right]^{-1} \qquad (11.12)$$

The calculation of the Bragg angle and the deviation parameter requires a rigorous procedure for dealing with strains and mismatches, whether caused by composition changes or, for example, ion implantation. A suitable method is as follows. All deviation angles are referred to the ideal Bragg angle of the substrate. The procedure for calculating the deviation parameter for an arbitrary layer is then:

1. Find the unit cell of the layer material in the fully relaxed (i.e., bulk) condition from materials databases. For alloys, Vegard's law[18] is applied to obtain the lattice parameters, Poisson ratio (ν), and structure factors. Find the susceptibility of the layers (the extremely small change in susceptibility when a layer is strained may be ignored).
2. Calculate the strains ε_{xx} and ε_{yy} that would be applied if the lattice parameters in the interface plane of the layer were forced to conform

to the substrate (full coherent epitaxy). Multiply these by $(1 - R)$, where R is the (fractional) relaxation of the layer (see Section 11.4.5).

3. Calculate the layer strain normal to the interface, ε_{zz}, from the relationship[19]

$$\varepsilon_{zz} = -(\varepsilon_{xx} + \varepsilon_{yy})\left\{\frac{\nu}{1-\nu}\right\} \tag{11.13}$$

4. Apply the strains to the layer material, giving a new unit cell. Hence, obtain the Bragg angle of the layer, the difference $\Delta\theta$ between the Bragg angle of the layer and that of the substrate, and the angle of tilt $\Delta\tau$ between the diffracting planes in the layer relative to those in the substrate. The deviation parameter α_h is then calculated separately for the different diffraction geometries, as follows:

Symmetric case:

$$\alpha_h = \Delta\theta_s - \Delta\theta \tag{11.14}$$

Asymmetric, glancing incidence:

$$\alpha_h = \Delta\theta_s - (\Delta\theta - \Delta\tau) \tag{11.15}$$

Asymmetric, glancing exit:

$$\alpha_h = \Delta\theta_s - (\Delta\theta + \Delta\tau) \tag{11.16}$$

where $\Delta\theta_s$, the deviation from Bragg angle in the substrate, is the controlled parameter, which forms the abscissa of the eventual rocking curve graph.

It is seen from the above equations that a tetragonal distortion is assumed in the layer. This is only strictly true if the substrate surface orientation is (001), though for symmetric reflections the treatment is valid for any orientation under the assumption of isotropic elasticity. However, the distortion would, for example, be trigonal on a (111), which would significantly affect the Bragg angle calculations.

The fitting of experimental high-resolution rocking curves and reflectivity profiles to data calculated from model structures is so important that we devote Chapter 13 to this topic. We do now have all the theoretical tools that we need to do the forward calculations. In the x-ray reflectivity (XRR) case, we use the Parratt formalism to calculate the specular reflectivity; in the case of high-resolution diffraction, we use the normalized intensity calculated from the Takagi–Taupin equations. The software-fitting engine performs these calculations many times as the model structure is iterated toward the optimum solution. However, in order to understand how the simulation routines work, it is beneficial to examine how some epilayer parameters can be extracted directly from the data. This provides a reliability check on both the final best-fit model and the quality of the experimental data themselves, and is an aid to planning a metrology recipe.

11.3 The Determination of Epilayer Parameters

In simple cases, parameters such as thickness, composition, tilt, and curvature can be extracted directly from peak positions and splittings on the rocking curve or $\omega - 2\theta$ curve (the latter is usually used in order to improve the signal/noise for weak signals). Relaxation requires more than one measurement. In laboratory development, software tools that extract these parameters directly from peak positions are often used. However, in a fab tool, consistency and repeatability of the analysis and of the parameters that are extracted is essential. Even in simple cases (such as a single uniform epilayer on a substrate) it is normal to perform the full analysis, by modeling of the structure and automated fitting to the full profile of the experimental data. This approach provides much better repeatability. It uses the theory of Section 11.2, and the modeling aspects are discussed in detail in Chapter 13.

11.3.1 Selection of Experimental Conditions

One of the most important parameters to control is the penetration depth. For this reason, most high-resolution diffractometry on semiconductors is done in reflection, since these techniques provide information only from the relevant surface region. In reflection, the intensity of the x-ray wave field inside the crystal falls off very rapidly away from the surface, due to transfer of energy to the diffracted beam. Absorption also becomes important at low incident angles to the surface. By choosing the radiation and the reflection (including its symmetry), the penetration may be varied between about 0.05 and 10 microns. This is ideally matched to device structures. This is quantified by the *extinction distance* ξ_g, defined as the depth at which the incident intensity has decreased to $1/e$ of its value at the surface. This may be calculated from diffraction theory, and some examples for GaAs and Si with CuKα radiation are shown in Table 11.1. It is assumed that the wafer surface is (001); hence, the 004 reflection is symmetric and the others asymmetric. The 004 is also the strongest accessible reflection.

The effect of absorption will be to decrease the effective extinction distances at the lower glancing angles. To get to really low extinction distances with characteristic radiation, it is possible to use skew reflections to get very low glancing or grazing angles (see the inset box in on p. 84).

TABLE 11.1

Extinction Distances with σ Polarized CuK Radiation

	Extinction Distance (μm)	
Reflection	**GaAs**	**Si**
004	4.7	10.5
044	8.5	18.7
115	10.3	22.3
224	6.0	14.8

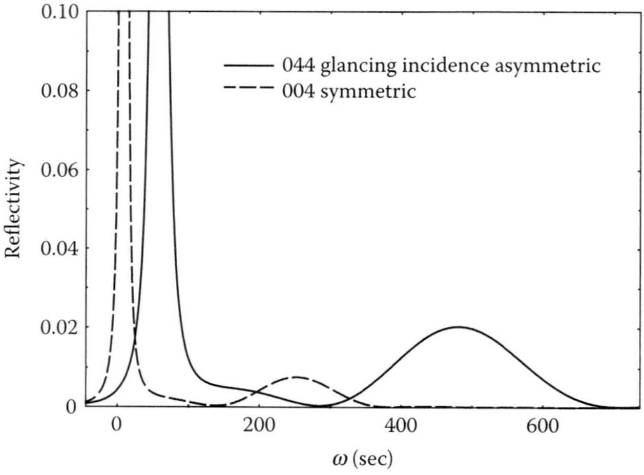

FIGURE 11.2
The effect of glancing incidence geometry, calculated for a 150-nm $In_{0.5177}Ga_{0.4823}As$ film on a (001) InP substrate. The symmetric 004 and the grazing incidence 044 curves are shown.

The minimum penetration in diffraction is obtained at or below the critical angle for total external reflection, when the penetration is then a few atomic layers. This gives information about surface layers. Figure 11.2 shows the dramatic effect obtained for a thin epilayer of InGaAs on InP. There is then a very rapid transition to external reflection.

Most of the important parameters may be derived from symmetric reflections, so there is usually no reason to depart from these, unless control of penetration depth is required. The exception is the measurement of relaxation, that is, the extent to which the interface is less than perfectly coherent with the substrate. A large mismatch will cause a substantial strain in the layer, which, at some critical thickness, will be relieved by the generation of interface dislocations. The average density of these may be accurately measured by measuring the lattice mismatch parallel as well as perpendicular to the interface. This requires an asymmetric reflection with a substantial component of the reflection vector parallel to the interface, in addition to the symmetric reflection and the determination of layer tilt relative to the substrate.

The accuracy and repeatability of all measurements will be improved if the background (away from the peaks) is low. This should never be more than a few cps if beams of 1 × 1 mm^2 or less are used, and <0.1 to 0.2 cps is attainable with standard instruments and a good detector. It is normally necessary to use a detector slit and perform $\omega - 2\theta$ scans rather than simple rocking curves (ω scans) to ensure good enough signal/noise to measure features in thin epilayers (see Chapter 3, Section 3.4.4).

We will henceforth assume that the actual parameter extraction will be performed by a modeling method set up by a specialist (if required, the equations for manual analysis can be found in Bowen and Tanner[20]). We shall discuss the effects on the experimental data of the parameters that can be extracted and the sensitivity of the data to variations in the parameter. Understanding of these issues is important for the design of efficient data acquisition.

There will in general be differences of diffraction angle between a layer and the substrate, caused either by tilt ($\delta\theta$) or mismatch (δd). Double or multiple peaks will therefore arise in the rocking curve. The substrate peak forms an important self-reference, so it is usual to speak of the splitting of the peak of a layer from that of the substrate. Peaks may be broadened by defects if these give additional rotations to the crystal lattice, and there will also be small peaks arising from interference between waves scattered from the interfaces, which will be controlled by the layer thicknesses. Finally, the material may show different defects in different regions. Table 11.2 summarizes qualitatively the influence on the rocking curve of the important parameters.

11.3.2 Measuring Composition

A difference in composition between substrate and layer results in a peak splitting $\delta\theta$. This is related to the change of interplanar spacing *normal* to the substrate through the equation

$$\delta d / d = -\delta\theta \cot\theta \tag{11.17}$$

If the reflection is the usual symmetric 004, then the experimental mismatch is

$$m^* = \delta a / a = \delta d / d \tag{11.18}$$

assuming, for the moment, that the layer is not tilted with respect to the substrate. This is derived directly from the measurements with no extra parameters or assumptions. However, this is not the mismatch that would

TABLE 11.2

The Effect of Substrate and Epilayer Parameters upon the Rocking Curve

Parameter	Effect on Rocking Curve	Distinguishing Features
Mismatch	Splitting of layer and substrate peak	Invariant with sample rotation
Misorientation	Splitting of layer and substrate peak	Changes sign with sample rotation; easily measured with triple-axis methods
Dislocation content	Broadens peak	Broadening invariant with beam size (if dislocation distribution is uniform across the beam size); no shift of peak with beam position on sample
Mosaic spread	Broadens peak	Broadening may increase with beam size, up to mosaic cell size; no shift of peak with beam position on sample
Curvature	Broadens and displaces peak	Broadening increases linearly with beam size; *w* position of peak shifts systematically with beam position on sample
Relaxation	Changes splitting from that of an unrelaxed layer	Different effect on symmetrical and asymmetrical reflections
Thickness	Affects intensity of peak; introduces interference fringes	Integrated intensity increases with layer thickness, up to a limit; fringe period controlled by thickness
Inhomogeneity	Effects vary with position on sample	Individual characteristics may be mapped

be measured if the layer were removed from the substrate and allowed to relax to its natural, unstressed state. The epilayer is elastically constrained to match the substrate in the plane *parallel* to the substrate, in both x and y directions, and there is a consequent *tetragonal distortion* of the epilayer, as shown in Figure 11.3.

The true or relaxed mismatch, m, is defined with respect to the relaxed (i.e., the normally tabulated) lattice parameters a_l and a_s of the epilayer and substrate materials as

$$m = (a_l - a_s) / a_s \tag{11.19}$$

It can then be calculated by means of some straightforward elasticity theory[21] as

$$m = m^* \left\{ \frac{1-\nu}{1+\nu} \right\} \tag{11.20}$$

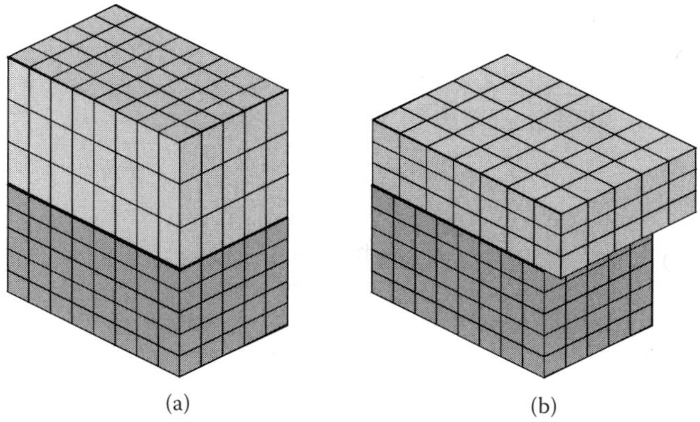

FIGURE 11.3
The tetragonal distortion in a coherent epilayer. (a) Fully relaxed. (b) Constrained to match the substrate.

where ν is the Poisson ratio. Since $\nu \sim 1/3$, $m \sim m^*/2$ as an approximate guide. For accurate work, the Poisson ratios of the constituents must be determined, and in principle these will vary with the alloy content.

For a binary or ternary layer the composition follows from Vegard's law; this simply states that the lattice parameter of a solid solution alloy will be given by a linear dependence of lattice parameter on composition, following a line drawn between the values for the pure constituents. Vegard's law[22] was originally proposed for ionic salt pairs, e.g., KCl-KBr, but has been widely assumed for semiconductors. It is based on elastic interactions between atoms, and is thus reasonable when all electronic interactions in the alloy series are very similar. The deviations from Vegard's law, which are usually of the order 1 or 2%, have been accurately calibrated for Si-Ge and Si-Ge-C alloys, which is of great importance in the silicon industry.

The intrinsic accuracy of mismatch measurements is excellent. For example, a 1-μm layer that is mismatched by 220 ppm from an InP substrate gives a peak splitting of 28"; this can be measured to about 2%, giving a resolution of 4 to 5 ppm.

In quaternary multilayers, only the mismatch may be obtained from the peak splitting. The composition cannot be obtained uniquely from the mismatch since a degree of freedom remains. A photoluminescent measurement of the band gap, or an XRF determination of composition ratios plus the mismatch measurement is needed.

11.3.3 Measuring Thickness

The layer thickness determines the relative intensity of the layer and substrate peaks. If the layer structure is simple, then the intensity of the layer peak increases monotonically with thickness. The values and the calibration constant will depend on the particular system being measured and can be determined either empirically or by computer calculation. The proper value to use for intensity measurements is the integrated intensity (the area under the peak) rather than the peak intensity. This is because it is less variable with material structure. If many dislocations are present, for example, in a strained layer material such as InGaAs on GaAs of moderate thickness, then the layer peak will be lowered and broadened, but to first order the integrated intensity will be the same. This method is little used in production control, since it normally does not give the required repeatability, but can be useful during initial development.

As discussed in Chapter 2, diffraction scans of epilayers show interference structures around the layer peak or peaks. These are called Pendellösung fringes (after the German word for *pendulum*, since they resemble the oscillations of a compound pendulum) and are the best way of measuring layer thickness. The most accurate way is through computer simulation (Chapter 13), but it is often useful to get a start to the simulation by measuring the interference peak separation, $\Delta\omega_p$, which is given by

$$\Delta\omega_p = \frac{\lambda\gamma_g}{t\,\sin 2\omega} \tag{11.21}$$

where λ is the wavelength, t the thickness, and γ_g the cosine of the angle between the diffracted beam and the inward-going surface normal. ω is accurately enough taken as the Bragg angle θ_B. For the reflection case, this may then be expressed as

$$\Delta\omega_p = \frac{\lambda\,\sin(\theta_B \pm \tau)}{t\,\sin 2\theta_B} \tag{11.22}$$

where τ is the angle between the reflecting plane and the surface; the positive sign applies to grazing incidence and the negative sign to grazing exit. For the common symmetrical case, we may simplify and rearrange to obtain

$$t = \frac{\lambda}{2\Delta\theta_p \cos\theta_B} \tag{11.23}$$

This very useful method also has the advantage that the equations do not contain anything about the material or diffraction conditions other than the Bragg angle and geometry. The independence from material parameters arises because the refractive index for x-rays is very close to unity. The equations are, of course, similar to those for optical interference from thin films, since the physics is the same, but in the optical case we do need to know the refractive index.

If more than one layer is present, there will be interference fringes from each of them. If their thicknesses are different, these will superimpose and beat. Very complex patterns may arise, but the modeling method distinguishes the contribution of different layers quite easily.

11.3.4 Measuring Tilt

If the layer is tilted relative to the substrate, then this will result in a shift of the layer peak relative to that of the substrate for reasons unconnected with composition. We have two problems: how to measure true splittings and how to determine the tilt itself.

The peak shift due to tilt will vary with the absolute direction of the incident beam relative to the substrate (i.e., with respect to rotation of the specimen about its surface normal), and may thereby be distinguished from mismatch splitting. If the specimen is rotated ϕ about its normal, the displacement of the layer peak from the position it would have were there no tilt is $\beta \cos\phi$, where β is the angle of tilt. It follows that a true splitting mismatch may be taken by rotating the specimen 180° in its plane and averaging the two measurements of splitting, i.e.,

$$\delta\omega = (\delta\omega_0 + \delta\omega_{180}) / 2 \tag{11.24}$$

This is insufficient to measure the tilt itself since we do not know whether our original measurement was in the direction in which the tilt would be a maximum (i.e., with the reflecting plane normals in the reference crystal, substrate, and layer all coplanar). A third measurement (at least) is required, followed by fitting the three measurements of layer peak deviation to a sine curve to find the maximum deviation — this is then the tilt value. The fitting is a simple computer iteration, which can provide accuracies of ~0.01 in the measurement of tilt, either of a wafer (offcut) or of a layer with respect to the wafer (Figure 11.4).

11.3.5 Measuring Curvature and Mosaic Spread

If the specimen is curved, then the change in angle from one side of the beam to the other will be cross-correlated with the rocking curve, and the latter will be broadened. Some information is lost; for example, fine interference fringes will be washed out. An estimate of this effect is straightforward. Let the radius of curvature of the specimen be R and the diameter of the beam be s. The angular change of the incident angle across the beam will then be s/R, as illustrated in Figure 11.5. For standard measurements, we want this to be a small fraction of the intrinsic rocking curve width; with III-V materials we might accept a broadening of 2" or 10^{-5} radians, which is 10 to 15% of the intrinsic width with CuKα radiation. If the beam diameter

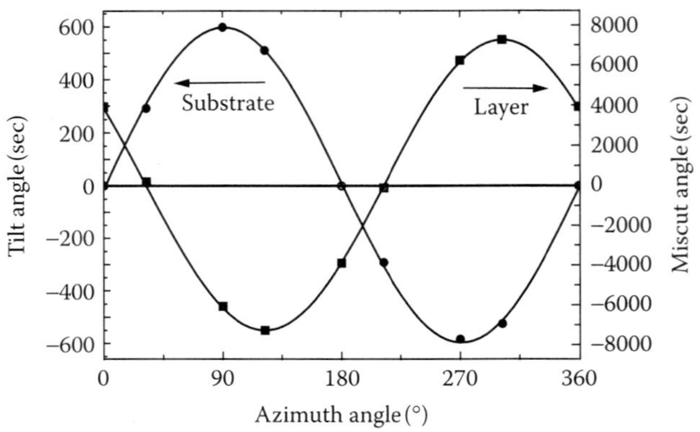

FIGURE 11.4
Determination of the tilt in a wafer and epilayer by measuring the peak position as a function of rotation about the surface normal.

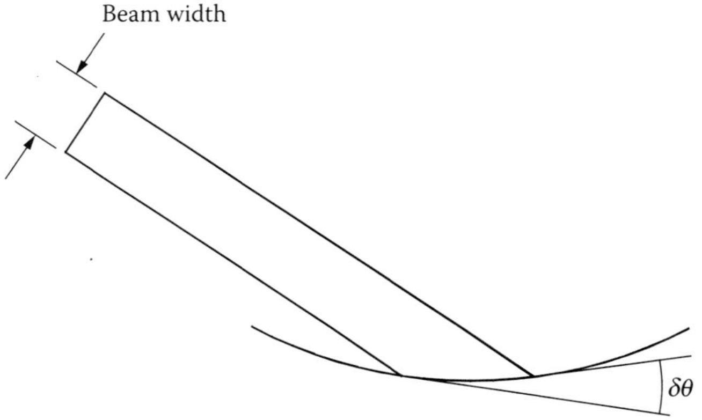

FIGURE 11.5
The change of incident angle across the beam on a curved specimen.

is 1 mm, then our criterion is satisfied if the specimen radius of curvature is no less than 100 m. This is not always true and emphasizes the need for small beams if accurate double-axis rocking curve widths are to be obtained on stressed specimens, though the splitting is not affected.

The curvature itself can be quite easily measured by translating the specimen a distance x in its plane along a diameter (a circular wafer is assumed), repeating the measurement and noting the shift $\delta\omega$ in the absolute position of the Bragg peak. Then, again,

$$R = s/\delta\omega \tag{11.25}$$

This measures the curvature about an axis perpendicular to the dispersion plane, i.e., the cylindrical curvature, and it may be necessary to rotate the

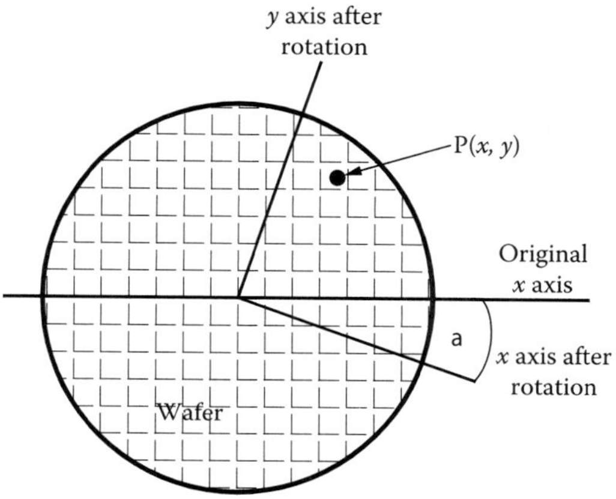

FIGURE 11.6
The projection of measured points onto the dispersion plane.

wafer through 90° to get the orthogonal component. This may be related to absolute stress in the wafer with knowledge of the wafer thickness, diameter, and elastic modulus. The most accurate method is to measure a number of points on a wafer and use a linear regression formula for the average curvature. If the specimen has been rotated (Figure 11.6), then we have

$$s = x\cos\phi + y\sin\phi \tag{11.26}$$

$$\omega_{substrate} = \left[\frac{1}{R}\right]s + \omega_0 \tag{11.27}$$

The linear regression (least squares) is performed on $\omega_{substrate}$ and s to get $1/R$. The residuals give the uniformity of curvature.

11.3.6 Measuring Dislocation Content

Dislocations are commonly present in three regions. A layer with high mismatch may relax so that interface dislocations are created to accommodate the strain. A network at the interface is thus observed. Slip dislocations may be generated by local plastic deformation due to thermal or mechanical strain and propagate elsewhere in the layer. Threading dislocations in the layer itself may also be generated during the growth process.

Interface dislocations give a specified relaxation of strain between the substrate and the epilayer, which gives quantifiable shifts in the positions of peaks in asymmetric reflections, as discussed later in this chapter. As structures become more complex, it is difficult to know which effects may be ascribed to interface relaxation and which to the layer structure itself. It is therefore often very useful to perform topography to see the dislocations directly, as discussed in Chapters 4 and 12.

On the other hand, dislocations inside the epilayer may have any of the possible Burgers vectors, and will on average contain roughly equal numbers of each sign. These do not shift the rocking curve, but they both broaden it and add diffuse scattering. A simple model for the broadening was given by Hirsch,[28] who showed that a reasonable estimate for the dislocation density ρ is

$$\rho = \frac{\beta^2}{9b^2} \tag{11.28}$$

in square centimeters, where β is the broadening of the rocking curve in radians and b is the Burgers vector in centimeters.

The diffuse scatter arises because dislocations are defects that rotate the lattice locally in either direction. This gives rise to scatter, from near-core regions, which is not traveling in quite the same direction as the diffraction from the bulk of the crystal. This adds kinematically (i.e., in intensity not amplitude) and gives a broad, shallow peak that must be centered on the Bragg peak of the dislocated layer or substrate since all the local rotations are centered on the lattice itself.

We can model the diffuse scatter quite well by a Lorentzian function of the form

$$I = \frac{A}{\Gamma^2 + \Delta\theta^2} \tag{11.29}$$

where A and Γ are constants. A further effect that can be seen in thick layers with large mismatch, for example, Ge or GaAs on Si, is the decrease of dislocation density from interface to surface; this is aimed for by the crystal growers and gives a rocking curve that is hard to simulate. However, an indication of the decrease in dislocation density can be obtained using two measurements: one a usual symmetric 004 and the other highly asymmetric to confine the beam to a region much nearer the surface.[24] If the dislocation content indeed decreases toward the surface, then rocking curves taken with smaller extinction distances (and the same area measured) will show less broadening.

11.3.7 Measuring Relaxation

Equation 11.20 contains the assumption that the interface is fully coherent. If it is only partially coherent, i.e., it contains interface dislocations, it is said to be relaxed, and Equation 11.20 is not valid for the determination of the relaxed mismatch. (Note the two different usages of the word *relaxed* in the

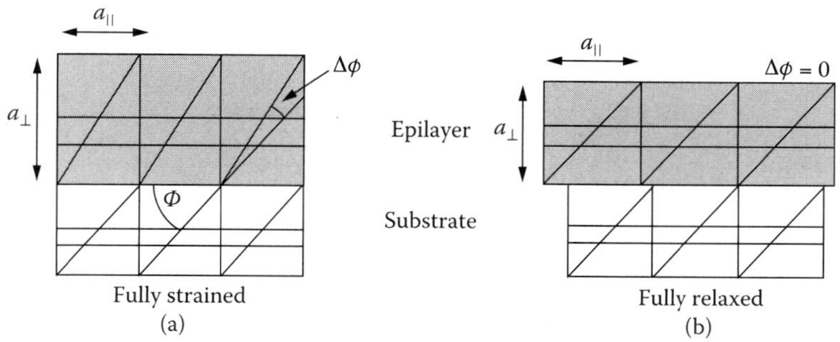

FIGURE 11.7
A side view of (a) coherent and (b) partially relaxed epilayers. The relaxation process changes both the interplanar spacings of the epilayer and the angles between the reflecting planes and the surface.

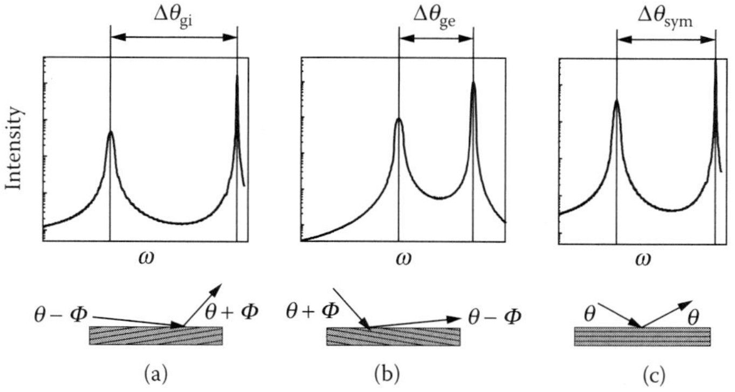

FIGURE 11.8
$\omega - 2\theta$ scans in (a) grazing incidence geometry, (b) grazing exit geometry, and (c) symmetric geometry. The differences in the peak splitting are evident.

last sentence.) It is necessary to measure the misfit parallel as well as perpendicular to the interface. For this, we need an asymmetric reflection that is at as high an angle to the surface as possible. 224, 511, and 311 are all acceptable.

Figure 11.7 shows a coherent and a relaxed layer, and it is clear that both the mismatch and the misorientation change between the substrate and the layer. The *tetragonal distortion* changes. From Figure 11.7 it can be seen that the effect of tilt on the splitting is reversed if the specimen is rotated by 180° about its surface normal, but the splitting due to the mismatch will not be affected by such a rotation. Thus, we may make grazing incidence or grazing exit measurements (Figure 11.8) to separate the tilt from the true splitting.

The resulting measured splittings, $\Delta\omega_i$ and $\Delta\omega_e$, are now different between these two geometries:

Grazing incidence:

$$\Delta\omega_i = \delta\omega + \delta\tau \tag{11.30}$$

Grazing exit:

$$\Delta\omega_i = \delta\omega - \delta\tau \tag{11.31}$$

Thus, we may determine $\delta\omega$ and $\delta\tau$ independently. We need to know the lattice parameter of the layers parallel and perpendicular to the substrate, i.e., the *in situ* cell parameters a_l, b_l, and c_l of the layers. From these we may calculate the relaxation and (together with the Poisson ratio) the fully relaxed lattice parameter of the epilayer. This last is the value that we need to use in Vegard's law to find the composition of the epilayer. Equations to calculate this manually are given in Bowen and Tanner,[20] but the calculation is in practice performed by computer iteration or full-curve simulation and matching.

If layers are thick, so that the diffracted signals are strong, it is easy to measure the symmetric and asymmetric reflections with an open detector. However, if the layers are typical of those found in commercial SiGe-on-Si wafers, it is necessary to have a detector slit and then do $\omega - n\theta$ scans, rather than simple rocking curves (ω scans), in order to have enough signal/noise. The problem is that n is only equal to 2 for the 004 reflection and for fully relaxed asymmetric reflections. For strained or partially relaxed layers, it is dependent on the degree of relaxation, which of course is what is to be measured.

Originally, the measurement was performed by multiple scans with variable n, which was very time-consuming. Matney and Ryan[25] have shown that accurate results may be obtained with just two scans. The first is a conventional symmetric scan. From the splitting seen in this scan, a functional relation between ω and 2θ may be calculated for any given reflection. Scanning the diffractometer along this direction gives a direct measure of the relaxation. As will be seen later in this chapter, the scan is along the Q_x direction in reciprocal space, at a fixed value of Q_z calculated from the 004 reflection. An example of this determination was shown in Chapter 1, Section 1.5 and Figure 1.12.

Once the parallel mismatch is determined, some information about average dislocation density in the interface may be obtained. It is not possible unambiguously to determine the types of dislocation present, since different types of dislocation may combine to give the same strain. However, the parallel mismatch is entirely due to dislocation content in (or very near) the interface, and thus

$$\frac{a_l - a_s}{a_s} = \frac{b}{s} \tag{11.32}$$

where b is the magnitude of the Burgers vector of the interface dislocations projected in the direction given by the intersection of the incidence plane with the interface, and s is their spacing in the same direction. If the nature of the dislocations is known, for example, from electron microscopy or x-ray topography, and two asymmetric reflections are taken, the overall dislocation density in the interface can be determined unambiguously.

11.4 High-Resolution Diffraction in Real and Reciprocal Space

11.4.1 Triple-Axis Scattering

A double-axis system, with or without a monochromator, uses an open detector and therefore integrates the scattering from the specimen over all angles within its aperture. While this is quick and convenient, it loses information; in particular, the scattering from bent or mosaic crystals occurs at different settings of the specimen crystal for a given d spacing, and details such as interference fringes or narrow peaks can be lost or blurred. In some cases, such as Figure 11.9, the width of the slit before the detector determines the apparent material quality. This ambiguity can be removed by analyzing the direction of the scattered x-rays from the crystal in the triple-axis geometry.

Choice of analyzer determines the resolution of the x-ray tool. For low resolution, a Soller slit consisting of a stack of closely spaced metal or glass

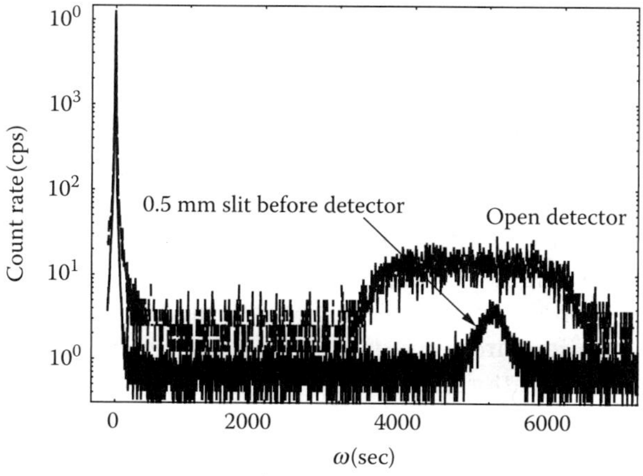

FIGURE 11.9
Double-axis rocking curves from a GaN epitaxial layer on (111) orientation GaAs. The measured rocking curve width is determined by the detector aperture.

foils, placed in front of the detector, restricts the angular acceptance of the detector while still preserving a large detector area. A typical Soller slit has an angular aperture of 0.1 to 0.15°. This is usually more effective than a single slit, as the beam footprint on the sample can be large without loss of resolution. Use of an auxiliary crystal in front of the detector, oriented to a selected Bragg angle, provides a method for analyzing the angular dependence of the scatter from the sample at much higher resolution. An asymmetrically cut Ge crystal set for the 220 reflection can provide an acceptance of up to 35 arc sec with CuKα1 radiation, while an asymmetric Si crystal under similar conditions will typically give up to 15 arc sec.

As just noted, the analyzer crystal is placed after the specimen and before the detector. It is mounted on an axis concentric with the specimen and is scanned independently of the sample. The x-ray metrologist can then map the intensity distribution with respect to the direction of the radiation scattered by the specimen, thus obtaining a *reciprocal space map* (also called a *triple-axis map*). This not only removes the complication of a possibly bent or mosaic specimen, but also enables one to distinguish scattering from various sources. For example, scattering due to defects occurs in a different direction in space than that from the perfect crystal. From a map of the scattering as both the specimen and analyzer are rotated, this can be measured quantitatively. Scattering from a rough surface can be separated from the perfect crystal scattering, and most importantly, strain or mismatch may be distinguished from tilt or mosaic spread.

11.4.2 Setting up a Triple-Axis Measurement

At first sight, the complexity of yet another Bragg reflection seems appalling. However, this is not the case with modern instruments with computer-controlled alignment. With a little experience, triple-axis measurements will be no more challenging than most double-axis ones. While the details of alignment will depend on the particular instrument, the tool is in all cases aligned when the maximum intensity is achieved after Bragg diffraction from monochromator, specimen, and analyzer crystals. Triple-axis measurements normally take this "everything at maximum" position as the zero setting for reciprocal space maps. Two qualifications are needed. The position will of course not be the origin of reciprocal space but the reciprocal lattice point corresponding to the reflection used. Second, in the usual Bragg case it will be displaced by a few arc seconds to higher angles by the refractive index effect in dynamical scattering.

11.4.3 Separation of Lattice Tilts and Strains

Triple-axis scattering enables the user to distinguish between tilts and dilations. This may be seen by considering the specimen shown in Figure 11.10, containing regions that are tilted with respect to each other, i.e., subgrains,

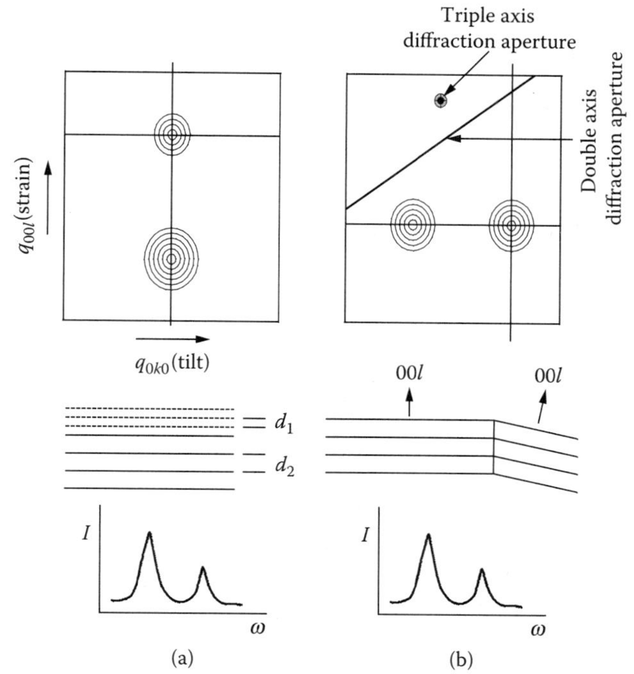

FIGURE 11.10
Triple-axis measurements: real and reciprocal space representations. In (a) the specimen has an epilayer, whereas in (b) there is no epilayer but a tilted mosaic region. The double-axis rocking curves cannot distinguish these and could even give similar rocking curves. There is no ambiguity in the reciprocal space maps.

and those that are strained or mismatched, e.g., ternary layers. By Bragg's law, the scattering angle defines the d spacing that is being examined. As the specimen is rotated, differently tilted regions will satisfy the condition for diffraction in sequence and the scattered intensity gives a measure of the distribution of tilts (see the inset box). Regions of the crystal where the Bragg plane spacing differs will never give rise to strong scattering when only the specimen is rotated.

Suppose we now perform a scattering experiment in which the specimen and analyzer are scanned in synchronization. Specifically, the analyzer is scanned at twice the rate of the specimen, both starting from zero. This is known as an $\omega - 2\theta$ scan, which may perhaps be better called an $\omega - \Phi$ scan. Suppose that a region of the specimen of lattice parameter d is set to diffract; then, as the analyzer is set at twice this angle, intensity reaches the detector. If we now perform an $\omega - \Phi$ coupled scan, any region of the specimen that also has lattice parameter d, but is tilted with respect to the original region, will never provide scattering that reaches the detector; the analyzer setting will never be correct (unlike the example in the previous section). However, another region of the sample with lattice parameter d' may come to a position where the Bragg angle is satisfied. Now the analyzer is at twice this angle

Triple-axis measurement of mosaic structure

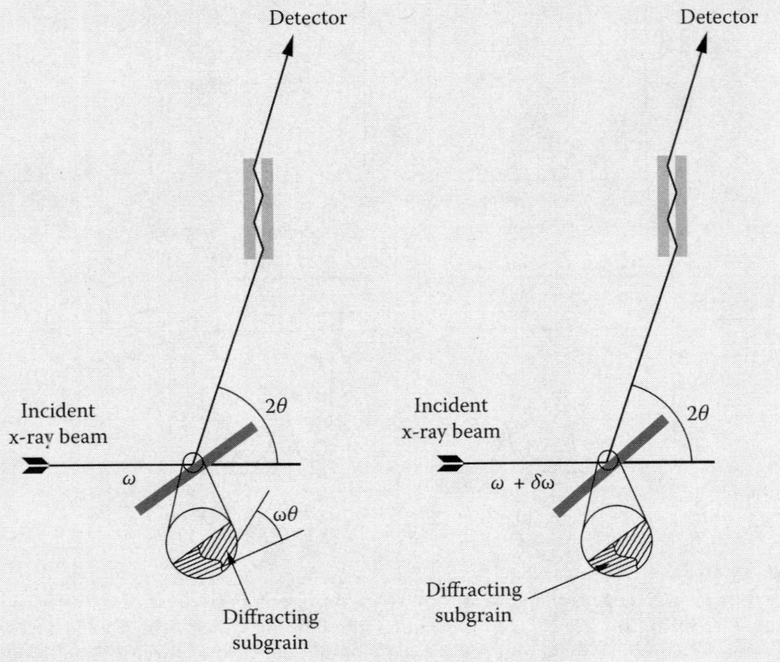

In the triple axis mode, the analyzer crystal passes only the diffraction from a single lattice parameter at a single orientation. Different mosaic domains thus diffract at different ω but the same 2θ setting.

and intensity reaches the detector. In this mode we record intensity from only parts of the crystal, but for heavily distorted materials such as gallium arsenide on silicon or small-gap II-VI compounds, this provides a valuable measure of the range of lattice parameters present. In the case of ternary compounds such as cadmium mercury telluride, this provides a measure of the composition range, independent of the range of tilts.

Note also the resolution functions in reciprocal space of the double- and triple-axis measurements. In the double-axis rocking curves (shown), we cannot distinguish between the tilts and the dilations, but they are very different types of defect. The triple-axis resolution function in the dispersion plane is the intersection of two vectors — incident and diffracted beam — spread out by the angular width of the monochromator and analyzer, respectively. A further blurring occurs in the plane normal to the figure, due to the vertical divergence, and the final component of the resolution function is the wavelength spread; this changes the lengths of the incident and diffracted beam vectors.

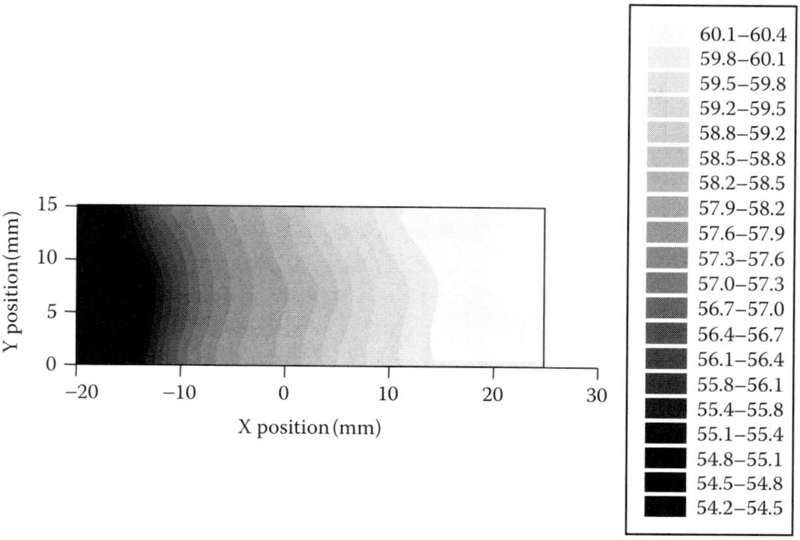

FIGURE 11.11
Variation in the indium composition (%) across an $In_xGa_{1-x}As$ substrate, determined from the Bragg peak position in a coupled $\omega - \Phi$ scan as a function of the location of the beam on the sample.

The scattering in the triple-axis geometry from different areas on a wafer can be combined to provide an independent view of the uniformity of the lattice spacing or the variation in tilt. Figure 11.11 shows an example of the variation in the indium content across an $In_xGa_{1-x}As$ substrate, determined from the Bragg peak position in a coupled $\omega - \Phi$ scan as a function of the location of the beam on the sample. Such information cannot be obtained in the double-axis rocking curve geometry (ω scan with open detector), as the presence of tilt across the wafer would also lead to an associated shift in the ω position of the Bragg peak, thereby making it impossible to distinguish the two effects.

11.4.4 Reciprocal Space Mapping

We have just alluded to the resolution in reciprocal space of the triple-axis setting. This representation permits us to map the scatter around a reciprocal lattice point and obtain a separate understanding of the origin of the diffuse and coherent scatter. Such a map of the scattering from the specimen can be made by recording intensity from a series of separate specimen and analyzer positions, which are coupled so as to trace out a grid in reciprocal space. Such a contour map is shown schematically in Figure 11.12, after Iida and Kohra.[26]

We may understand what is being measured in a triple-axis experiment with the aid of this figure. The angular positions of the incident beam and

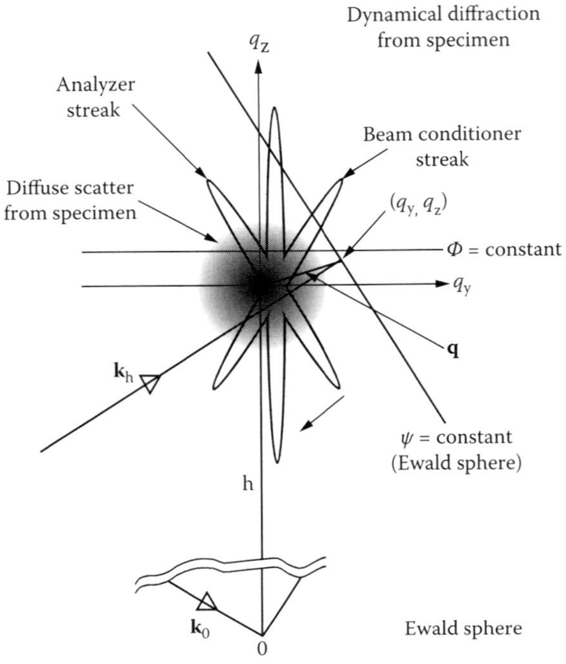

FIGURE 11.12
A scattering map in reciprocal space. Equal intensity contours are shown schematically, and the Ewald sphere is represented as a plane near reciprocal lattice points 0 and h. The dynamical diffraction from the specimen is displaced slightly from the reciprocal lattice point and from the center of the diffuse scatter by the refractive index effect. (Modified from Iida, A. and Kohra, K., *Phys. Stat. Sol.*, 51, 533–542, 1979.)

of the *analyzer* define two vectors that define the scattering vector and hence angle. The angular position of the specimen defines the position of the diffracting planes whose scattering is being measured at this scattering angle. In Figure 11.12, the central point is the reciprocal space point of the diffracting planes, for example, 004. The scattering is being measured in this instance from the small volume surrounding the point (q_x, q_z). The scattering vector **Q** (not shown directly in Figure 11.12) may be considered the sum of the ideal scattering vector from the origin to point h, plus a deviation **q**; thus,

$$\mathbf{Q} = \mathbf{h} + \mathbf{q} \tag{11.33}$$

The deviation vector **q**, with origin at the end of the reciprocal lattice vector **h**, has two components: q_x horizontal, positive rightward going, and q_z vertical, positive upward going. For the symmetric reflection, these components are related to the deviations of specimen ($\Delta\omega$) and analyzer ($\Delta\Phi$) from their zero positions at the nominal Bragg angle by the equations

$$q_z = \Delta\Phi \cos\theta_B / \lambda \tag{11.34}$$

$$q_x = (2\Delta\omega - \Delta\Phi)\sin\theta_B / \lambda \tag{11.35}$$

Thus, a scan of the specimen axis alone ($\Delta\omega$) affects only q_x and provides a scan from left to right in reciprocal space. A scan of the *analyzer* affects both q_z and q_x and, in fact, sweeps along the Ewald sphere. A scan of q_z alone may be achieved by setting

$$(2\Delta\omega - \Delta\Phi) = 0 \tag{11.36}$$

i.e., scanning the analyzer at twice the rate of the specimen, the so-called $\omega - 2\theta$ or $\omega - \Phi$ scan.

Geometrically, we may visualize the above scans as follows. Scanning the specimen alone is equivalent to rotating the reciprocal lattice about its origin; the end of $\mathbf{k}_h$ thus describes an arc about O, which at the scale of the drawing is a straight horizontal line. Scanning the detector alone is equivalent to changing the angle between $\mathbf{k}_o$ and $\mathbf{k}_h$, thus describing a scan along the Ewald sphere. The $2\Delta\omega - \Delta\Phi$ scan to move vertically in reciprocal space arises because $\Delta\psi$ and $\Delta\phi$ are respectively the angles standing on the center and the circumference of the Ewald sphere, subtended from the same arc; hence, $2\Delta\omega = \Delta\Phi$ by simple geometry of a circle.

The general case of the asymmetric reflection is a little more complicated and not usually given; the following derivation is due to Wormington.[27] The deviation of the scattering vector, **q**, from the reciprocal lattice point, **h**, can be calculated by considering the Ewald constructions shown in Figure 11.13. Figure 11.13a illustrates the case when the specimen angle, ω, and the analyzer angle, Φ, are set so as to satisfy the Bragg condition, $2d \sin\theta_B = \lambda$. Figure 11.13b shows the Ewald construction after the specimen and analyzer angles have been changed by $\Delta\omega$ and $\Delta\Phi$, respectively. On rearranging Equation 11.32 we have

$$\mathbf{q} = \mathbf{h} - \mathbf{Q} \tag{11.37}$$

From the geometry of the Ewald construction shown in Figure 11.13 we can write $\Delta\mathbf{q}$ in terms of its Cartesian components, (q_y, q_z):

$$q_x = \frac{1}{\lambda}\{\cos(\Phi + \Delta\Phi - (\omega + \Delta\omega)) - \cos(\omega + \Delta\omega)\} - \frac{1}{\lambda}\{\cos(\Phi - \omega) - \cos(\omega)\} \tag{11.38}$$

and

$$q_z = \frac{1}{\lambda}\{\sin(\omega + \Delta\omega) + \sin(\Phi + \Delta\Phi - (\omega + \Delta\omega))\} - \frac{1}{\lambda}\{\sin(\omega) + \sin(\Phi - \omega)\} \tag{11.39}$$

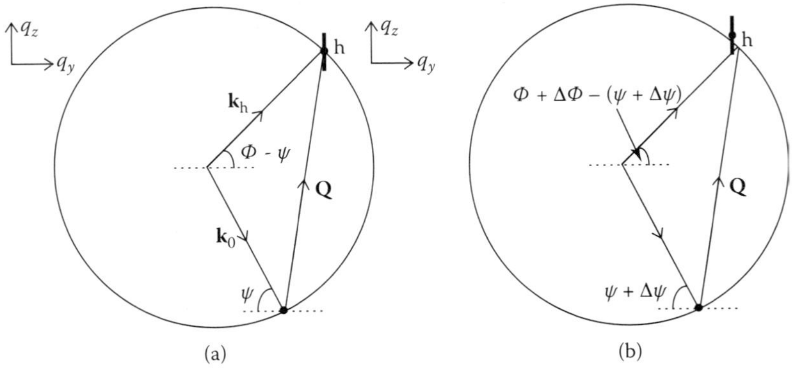

FIGURE 11.13
Ewald constructions (a) at the Bragg condition and (b) off the Bragg condition.

Expanding the trigonometric terms we may rewrite Equations 11.36 and 11.37 as

$$q_x = \frac{1}{\lambda}\{[\cos(\Phi - \omega)\cos(\Delta\Phi - \Delta\omega) - \sin(\Phi - \omega)\sin(\Delta\Phi - \Delta\omega)] - [\cos(\omega)\cos(\Delta\omega) - \sin(\omega)\sin(\Delta\omega)]\} - \frac{1}{\lambda}\{\cos(\Phi - \omega) - \cos(\omega)\} \tag{11.40}$$

$$q_z = \frac{1}{\lambda}\{[\sin(\omega)\cos(\Delta\omega) + \cos(\omega)\sin(\Delta\omega)] + [\sin(\Phi - \omega)\cos(\Delta\Phi - \Delta\omega) + \cos(\Phi - \omega)\sin(\Delta\Phi - \Delta\omega)]\} - \frac{1}{\lambda}\{\sin(\omega) + \sin(\Phi - \omega)\} \tag{11.41}$$

If $\Delta\omega$ and $\Delta\Phi$ are small, Equations 11.38 and 11.39 can be simplified by using the small angle approximations, $\sin(\Delta\omega) = \Delta\omega$ and $\cos(\Delta\omega) \approx 1$, and similarly for $\Delta\Phi$. We may finally write

$$q_x \approx \frac{1}{\lambda}\{\sin(\omega)\Delta\omega - \sin(\Phi - \omega)(\Delta\Phi - \Delta\omega)\} \tag{11.42}$$

$$q_z \approx \frac{1}{\lambda}\{\cos(\omega)\Delta\omega + \cos(\Phi - \omega)(\Delta\Phi - \Delta\omega)\} \tag{11.43}$$

where for symmetric reflections:

$$\omega = \theta_B,\ \Phi - \omega = \theta_B$$

for asymmetric reflections with glancing incidence:

$$\omega = \theta_B - \tau,\ \Phi - \omega = \theta_B + \tau$$

and for asymmetric reflections with glancing exit:

$$\omega = \theta_B + \tau,\ \Phi - \omega = \theta_B - \tau$$

τ is the magnitude of the angle between the diffracting planes and the surface normal.*

Several points arise from Figure 11.12. The main dynamical scattering from the specimen is a vertical streak, along the q_z direction, which in a good-quality crystal has little extension in the q_x direction. This scattering is displaced from the reciprocal lattice point by the refractive index correction. However, diffuse scattering appears as a weak but broad, approximately annular scattering region centered on the reciprocal lattice point.

Streaks appear at $\pm\theta_B$ to the vertical axis. These are caused by the finite angular resolution of the beam conditioner and analyzer crystals and would be absent if these crystals had rocking curves with no tails. They arise because at these settings of analyzer and specimen, in addition to the true signal from the sampled region of reciprocal space, a small proportion of the intense signals from the main peak pass through the tails of the analyzer function. Hence, the streaks point toward the main peak, and their direction is found by imagining a small oscillation of $\mathbf{k}_0$ about the origin and of $\mathbf{k}_h$ about the center of the Ewald sphere. In a system with high resolution and very wide dynamic range, similar streaks at very low intensity levels can sometimes be seen from air scatter in the system — this has a similar effect in blurring the angular precision.[28] Finally, we note that the standard double crystal or high-resolution rocking curve, in which we have no control over $\Delta\phi$, is in effect a horizontal scan through reciprocal space, integrating all intensities along the Ewald sphere. It is thus easy to see how the triple-axis instrument can obtain much more information.

11.4.5 The Relaxation Scan

Despite its visual impact and aid to understanding, collection of a reciprocal space map is time-consuming. For x-ray metrology in a semiconductor fab, a much more rapid data collection strategy is necessary. This involves

* This quantity is often notated as ϕ in other texts; we avoid this since it leads to confusion with the scattering angle Φ and the rotation angle about the specimen normal, ϕ.

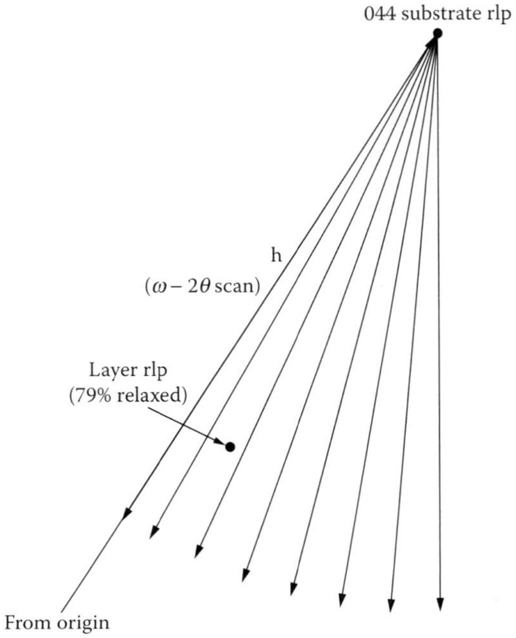

FIGURE 11.14
A section of reciprocal space around the 044 reflection. The set of scans that would find a relaxed layer rlp (reciprocal lattice point) is shown.

collecting single scans along specific directions in reciprocal space. The most obvious choice of scans are:

- A scan parallel to the local **h** vector, which is achieved by a coupled $\omega - \Phi$ (also called $\omega - 2\theta$) scan. For the symmetrical $00l$ reflection only, this is parallel to Q_z.
- A scan perpendicular to the local **h** vector, which is an ω (specimen) scan at fixed detector position. For the symmetrical $00l$ reflection only, this is parallel to Q_x.

The first of these is valuable for measuring strain independent of tilt, and the second for measuring tilt independent of strain. However, for measurement of the degree of relaxation of an epilayer, which requires an asymmetric reflection, these are both inappropriate since they are likely to miss the epilayer peak unless it is fully relaxed. The conventional way to find the layer peak was to perform a series of coupled $\omega - n\Phi$ scans, with n running from 1 down to the value (which depends on the reflection) that gives a scan parallel to Q_z. The partially relaxed peak would then be found on one of these scans (Figure 11.14). This is obviously inefficient.

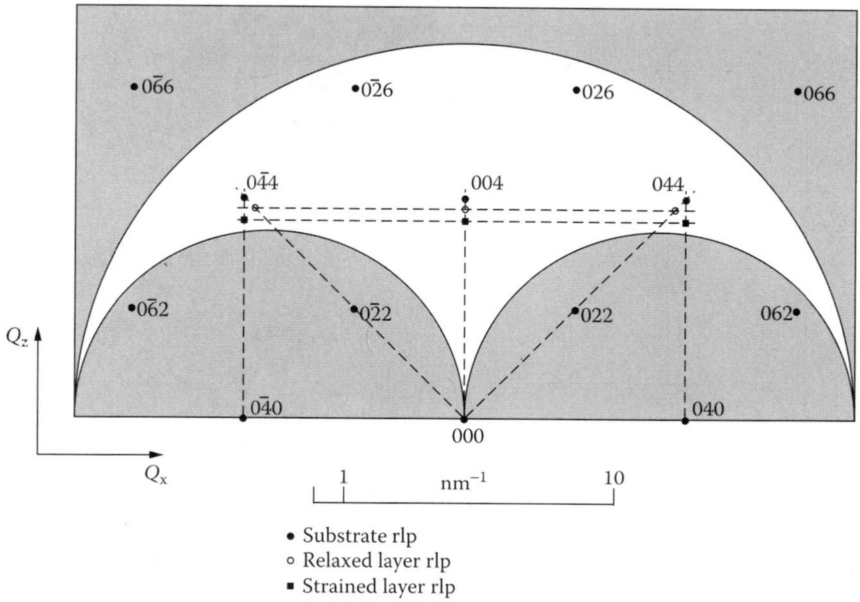

FIGURE 11.15
The reciprocal space map introduced in Chapter 10, showing (on exaggerated scales) the positions of the substrate and layer peaks for both unrelaxed and relaxed layers.

As introduced in Chapter 1, Matney and Ryan[25] developed a much more efficient method, requiring only two scans. The result was shown in Figure 1.12, and now we consider the theory behind the method. Figure 11.15 shows the full reciprocal space map, and the 004 and 044 reciprocal lattice points (rlps) are considered for the substrate and the layer, for a case such as SiGe on Si, where the unit cell of the relaxed layer is larger than that of the substrate. For the 004 reflection, the relaxed and strained layer rlps both lie on the Q_z axis, and thus cannot be distinguished. This is just another way of saying that we need a component of the measurement parallel to the interface in order to determine the lattice parameter in the interface. However, the relaxed and unrelaxed rlps lie on different Q_x coordinates in the 044 reflection. Furthermore, the Q_z value of the layer is the same in both reflections, whether it be relaxed or unrelaxed. This is because the Q_z component measures the component of lattice space resolved normal to the interface. Thus, from an 004 measurement, we know the line along which the layer peak must lie in its 044 reflection, in this case parallel to Q_x and at the same Q_z coordinate as was found in the 004 scan. This is illustrated schematically in Figure 11.16a and experimentally in Figure 11.16b.

For the case of reflections other than 044, the scan will still be along Q_x, but the z displacement from the substrate rlp will be in the ratios of the z components of the two reciprocal lattice vectors.

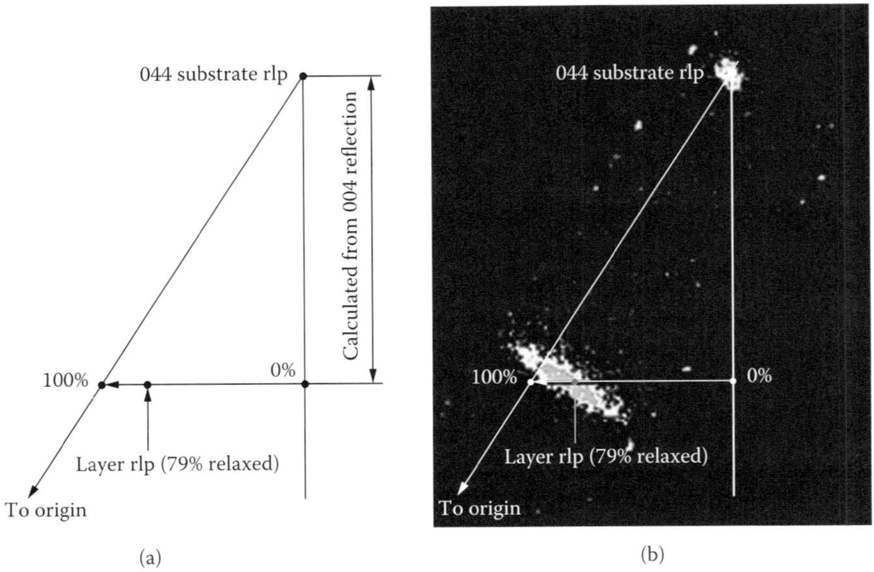

FIGURE 11.16
(a) Enlargement of the region around the 044 rlp, showing the relationship between the fully relaxed and fully strained layer peaks and the relaxation scan. (b) Experimental example on a 79% Si-Ge layer, shown at the same scale as in (a).

11.4.6 Grazing Incidence In-Plane Diffraction

The high-resolution x-ray diffraction (HRXRD) methods described above all probe the structure of atom planes either parallel to the wafer surface or inclined at an angle to it. Measurement from the lattice planes that are perpendicular to the wafer surface (i.e., at the special angle of 90°) cannot be done in the standard HRXRD geometry. However, in the scattering geometry shown in Figure 11.17, an unusual situation occurs. When the incident beam makes an angle of less than the critical angle with respect to the surface, the x-ray wave penetrates only to the depth of a few nanometers. If the crystal is oriented such that the Bragg angle is satisfied, corresponding to the planes normal to the surface, a diffracted beam is observed emerging at the same small angle to the surface as that of the incidence beam. Unlike the case of a polycrystalline film (Chapter 4), where appropriately oriented crystallites are self-selected for scattering, the exact Bragg condition cannot ever be satisfied for a single crystal. However, due to the presence of the surface, even for a very thick crystal, the region over which strong scattering is observed is extended along the normal to the surface. In reciprocal space notation (see Chapter 10), this corresponds to the reciprocal lattice point being streaked out into a rod normal to the surface. Thus, even though the exact Bragg condition is not met, a peak is observed as the scattering vector sweeps through this extended truncation rod. Because the exact Bragg

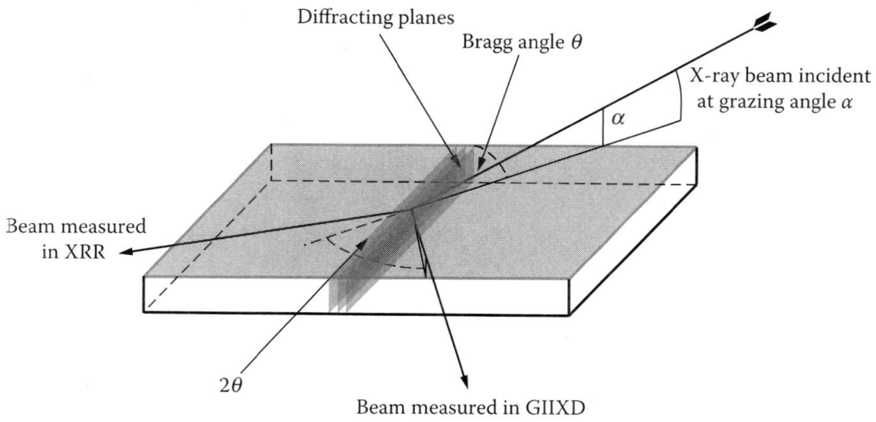

FIGURE 11.17
Schematic diagram of the GIIXD geometry, showing the incident and diffracted beams and the specularly reflected beam that is measured in standard grazing incidence reflectivity. (From B.K. Tanner et al., *Powder Diffraction*, 19, 45, 2004. An Investigation of Giant Magnetoresistance (GMR) Spin-Valve Structures Using X-ray Diffraction and Reflectivity, E. Brown and M. Wormington, *Advances in X-ray Analysis*, Vol. 44, © ICDD 2001. This material is used by permission of ICDD.)

condition is not met, the scatter from single crystals is weak. It is for this reason that, until recently, GIIXD has not been performed significantly in the laboratory. (There is, however, a large literature associated with such experiments at synchrotron radiation sources, particularly concerning the reconstruction of surface layers.) Much of the laboratory GIIXD work reported has been concerned with the measurement of the in-plane, so-called twist mosaic in highly mismatched epilayers.

To record GIIXD data with a normal x-ray tube, focusing optics are necessary together with long counting times and meticulous attention to reduction of background noise from air scatter and fluorescence. It can be performed either with a high-brilliance microfocus tube and polycapillary or total reflection optics or, alternatively, with a standard source and curved multilayer total reflection optics. Very high resolution in GIIXD can be achieved by use of an asymmetric Ge crystal to condition the incident beam, thereby reducing the beam divergence from the 0.15° or so that is provided by the x-ray optical components to typically 0.01° (Figure 11.18). (Use of a multilayer optic gives typically 0.05° incident beam divergence.) As always, the higher resolution is achieved at the expense of intensity.

The methodology for experiment setup is exactly the same as for HRXRD. Using an open detector, the sample is scanned, this time about the surface normal at a fixed grazing incidence angle until a Bragg peak is located. In the case of (001)-cut Si, there will be four such peaks separated by 90° (Figure 11.19). After location of the specimen, an analyzer consisting of either Soller slits or a crystal of graphite or Ge is introduced before the detector. Rocking curves can be performed to measure the mosaic distribution, while $\phi - 2\theta$

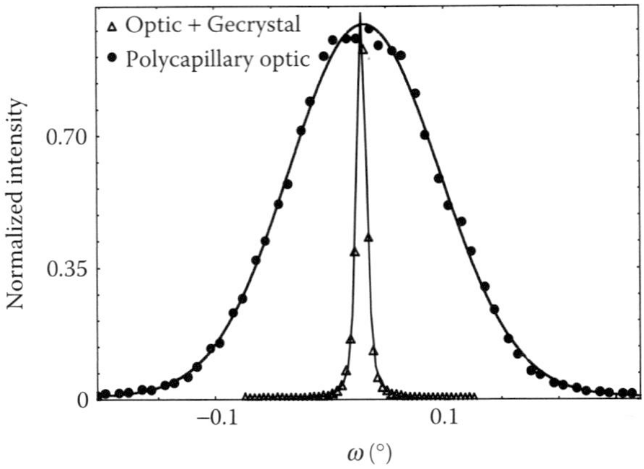

FIGURE 11.18
Rocking curves from a single crystal of silicon with a polycapillary optic only in incidence beam and also with an asymmetrically cut Ge (004) crystal included as an additional beam conditioner. (From T.A. Lafford et al., *Phys. Stat. Sol.*, 195, 265, 2003. With permission.)

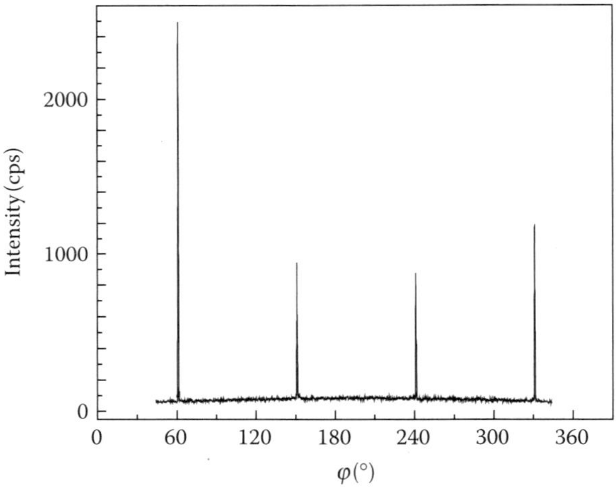

FIGURE 11.19
Scans of single crystal of Si about its surface normal with an open detector showing the symmetry of the GIIXD peaks. There are two peaks under each of the four peaks apparent in this figure. (From M.S. Goorsky and B.K. Tanner, *Cryst. Res. Technol.*, 37, 647, 2002. With permission.)

scans can be used to measure the absolute lattice parameter. As indicated earlier, the beam penetrates only a small depth into the surface, and hence the background signal from the substrate can be suppressed, giving only scatter from the thin surface layers of interest. (Note, however, that to achieve the few-nanometer-depth penetration associated with total external reflection, the beam divergence must be strictly limited in the direction normal to the surface. A divergence of 0.15° from a polycapillary optic is insufficient in most cases.) The technique thus has considerable potential for the study of strained silicon layers used in the manufacture of very fast silicon devices.

11.5 Summary

- High-resolution x-ray diffraction may be used to measure the out-of-plane lattice parameter, and with asymmetric reflections, the lattice parameter in the plane can be deduced.
- A combination of such HRXRD measurements can be used to determine the composition and relaxation independently.
- HRXRD basically measures strain and rotation in each epilayer. HRXRD can be used to measure composition, mosaic spread, tilt between substrate and epilayer, and wafer curvature, and to deduce an approximate, model-dependent value of dislocation density.
- Measurement of layer thickness is highly accurate, since it is an interferometric method using the very short wavelength of x-rays and does not depend upon uncertain optical constants.
- Reciprocal space mapping provides a valuable pictorial description of the scatter in reciprocal space that enables the origin of the scatter to be deduced.
- GIIXD provides a powerful method for direct measurement of the in-plane lattice parameter (and hence relaxation) together with the twist mosaic.

References

1. R.W. James, *The Optical Principles of the Diffraction of X-Rays*, Ox Bow Press, Woodbridge, CT, 1982.
2. A. Authier, *Adv. Struct. Res. Diffraction Methods*, 3 (1970) 1.
3. A. Authier, in *X-Ray and Neutron Dynamical Diffraction: Theory and Applications*, A. Authier, S. Lagomarsino, and B.K. Tanner, Eds., Plenum Press, New York, 1997, p. 1.
4. B.W. Batterman and H. Cole, *Rev. Mod. Phys.*, 36 (1966) 681.
5. M. Hart, in *Characterization of Crystal Growth Defects by X-Ray Methods*, B.K. Tanner and D.K. Bowen, Eds., Plenum Press, New York, 1980, p. 216.
6. Z.G. Pinsker, *Dynamical Scattering of X-Rays in Crystals*, Springer, Berlin, 1977.
7. W.H. Zachariasen, *Theory of X-Ray Diffraction in Crystals*, Wiley-Dover, New York, 1945 (reprint).
8. M. Hart, *Rep. Prog. Phys.*, 34 (1971) 435.

9. M. von Laue Rontgenstrahlinterferenzen, Akademie-Verlag, Verlag, Frankfurt, 1960.
10. A. Authier, *Dynamical Theory of X-Ray Diffraction*, Oxford University Press, Oxford, 2001.
11. U. Pietsch, V. HolΔ, and T. Baumbach, *High-Resolution X-Ray Scattering*, 2nd ed. Springer, New York, 2004.
12. S. Takagi, *Acta Cryst.*, 15 (1962) 1311.
13. S. Takagi, *J. Phys. Soc. Jpn.*, 26 (1969) 1239.
14. D. Taupin, *Bull. Soc. Fr. Miner. Cristallogr.*, 87 (1964) 469.
15. M.A.G. Halliwell, J. Juler, and A.G. Norman, *Inst. Phys. Conf. Ser.*, 67 (1983) 365.
16. Z.G. Pinsker, *Dynamical Scattering of X-rays*, Springer-Verlag, Berlin, 1978.
17. R. Zaus, *J. Appl. Cryst.*, 26 (1993) 801.
18. L. Vegard, *Z. Physik*, 5 (1921) 17.
19. L.D. Landau and E.M. Lifshitz, *Elasticity*, Pergamon Press, New York, 1972, pp. 13, 14, 55.
20. D.K. Bowen and B.K. Tanner, *High-Resolution X-Ray Diffractometry and Topography*, Taylor & Francis, London, 1998.
21. L.D. Landau and E.M. Lifschitz, *Elasticity*, Pergamon Press, New York, 1972.
22. L. Vegard, *Z. Physik*, 5 (1921) 17.
23. P.B. Hirsch, Mosaic structure, in *Progress in Metal Physics*, B. Chalmers and R. King, Eds., Pergamon Press, New York, 1956, chap. 6.
24. J.W. Lee, D.K. Bowen, and J.P. Salerno, *Mat. Res. Soc. Symp. Proc.*, 91 (1987) 193–198.
25. K.M. Matney and P.A. Ryan, private communication, 2003.
26. A. Iida and K. Kohra, *Phys. Stat. Sol.*, 51 (1979) 533–542.
27. M. Wormington, private communication, 1997.
28. R. Matyi, *Rev. Sci. Instrum.*, 63 (1992) 55.

12

Diffraction Imaging and Defect Mapping

12.1 Introduction

This chapter is concerned with mapping the type and density of defects on the wafer. In such measurements, the intensity variation across the diffracted beam is recorded, and thus a map of the scattering power is recorded as a function of position. The resulting images were traditionally called x-ray topographs, and they are analogous to transmission electron micrographs. Despite the name, the technique is not principally sensitive to surface topography; it is the topography of the crystal lattice planes that is examined. We shall use the more descriptive term x-ray diffraction imaging, except when we refer to explicit historical work.

Until recently, good-quality images could only be obtained by high-resolution photographic film or plate. These are quite unsuitable for use in a fabrication line, a fact that inhibited use of XRDI even for problems for which it is the best possible technique. However, fully digital systems are now available. These use either image plates at relatively poor resolution with conventional methods, or high-resolution image detectors with the new "virtual scan" (BedeScan™) method introduced in Chapter 1.6 and the inset on p. 26. Example images were shown in Figure 1.13.

12.2 Contrast in X-ray Diffraction Imaging (XRDI)

The aim of all x-ray diffraction imaging methods is to provide a picture of the distribution of the defects in a crystal, and the x-ray images may be thought of as arising in two ways. Orientation contrast arises when a region of the crystal is misoriented by an amount larger than the beam divergence (Figure 12.1). Then for the characteristic line, no diffracted intensity is recorded for region B when the Bragg condition is satisfied for region A. Thus, there is an unexposed patch on the detector. The angle of misorientation,

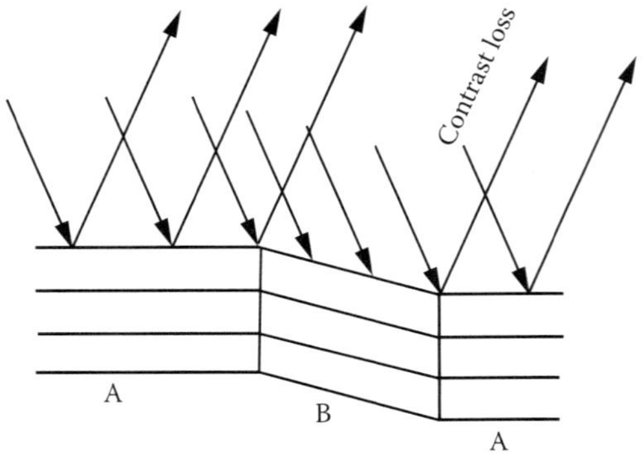

FIGURE 12.1
Orientation contrast from a monochromatic, collimated x-ray beam. Diffraction is missing in region B when the specimen is set to diffract from region A.

projected into the incidence plane, can be determined by the angle that the specimen must be rotated in order to obtain strong intensity from region B. This is very conveniently achieved in real time using an electronic imaging detector, but is tedious to perform with photographic recording. Orientation contrast may arise from the presence of twins, subgrains, and electric and magnetic domains, and can be interpreted in a simple geometric manner.

An example from conventional Lang topography is shown in Figure 12.2. The specimen is a single crystal of $TbAlO_3$ grown from a fluxed melt. As well as the dislocations marked **d** and faint bands **p** associated with fluctuations in impurity level, the dominant feature is a vertical stripe marked **A**. It is bounded on each side by narrow twins, which results in the lattice being misoriented by an amount much larger than the beam divergence. Therefore, in topograph (a) no intensity arises from the twin-bounded region, through the mechanism of orientation contrast. In topograph (b) the diffraction vector is changed to be parallel to the axis of rotation of the twins, and the region is in good contrast. Optical micrograph (c) shows the narrow twins.

This discussion also illustrates a general feature of diffraction imaging contrast. If the distortions in a crystal are all parallel to the diffracting plane, then they do not give rise to image contrast. Therefore, we may use different reflections to map out the orientation quantitatively, for example, to determine the Burgers vectors of dislocations.

Another type of orientation contrast arises in the BedeScan method and also when continuous radiation is being used, for example, at synchrotron sources. Here the two regions A and B satisfy the Bragg condition at different angles. With continuous radiation this is because they arise from different wavelengths; with the BedeScan method the incident beam tracks the orientation change. The exit beams from the rotated region then make different directions in space with those from the neighboring regions. They may either overlap or diverge, depending on the relative misorientation of regions A

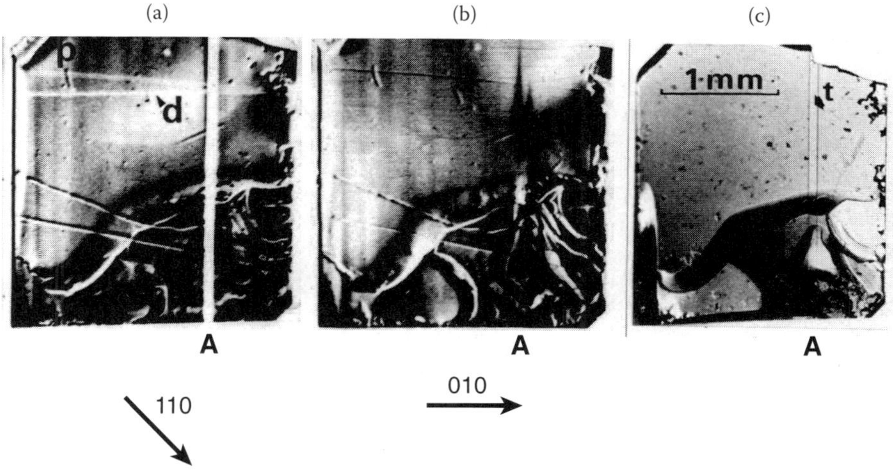

FIGURE 12.2
An example of orientation contrast in Lang topography. (a) X-ray Lang topograph (diffraction image in transmission) of a 2-mm-wide single crystal of $TbAlO_3$ grown from a fluxed melt. The diffraction vector direction [110] is indicated. (b) As (a), but with horizontal diffraction vector [010]. (c) Optical polarized light micrograph. (From Wanklyn, B.M. et al., *J. Crystal Growth*, 29, 281, 1975. With permission from Elsevier.)

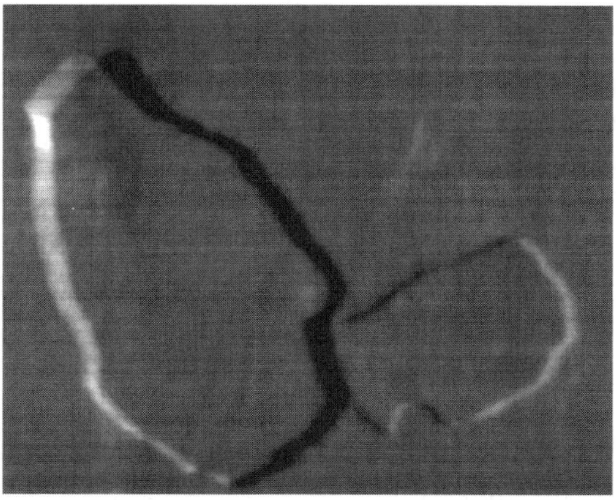

FIGURE 12.3
An example of orientation contrast from two subgrains in a lithium niobate crystal. BedeScan image in reflection, CuKα, 115 in reflection.

and B. Simple images occur where the crystal contains discrete mosaic blocks or subgrains (for example, Figure 12.3), but when the lattice distortion is continuous, the resulting contrast can be very complex and difficult to interpret.

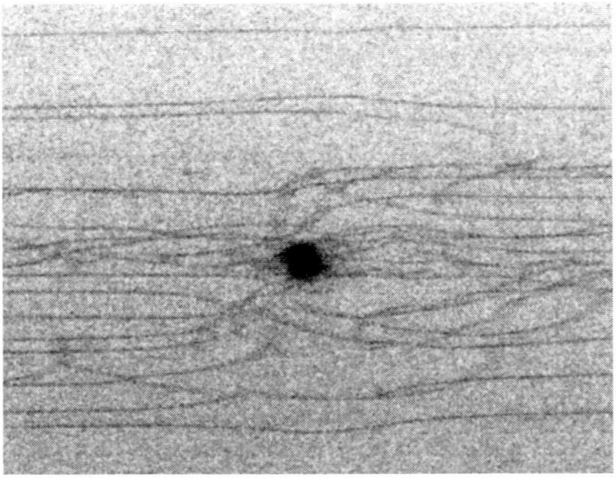

FIGURE 12.4
Individual dislocations arising from a surface strain point in a silicon wafer. BedeScan image in transmission, 220 in reflection.

The second major contrast mechanism is extinction contrast. Here the distortion of the lattice around a defect gives rise to a different scattering power from that of the surrounding matrix. In all cases, it arises from a breakdown or change of the dynamical diffraction in the perfect crystal. In classical x-ray structure analysis, the name *extinction* was used to describe the observation that the integrated scattered intensity was less for a perfect crystal than that predicted by the kinematical theory. Around the defect, enhanced (kinematical) scattering was observed and this loss of extinction is the origin of the name. An example of extinction contrast from individual dislocations is shown in Figure 12.4.

Since we are dealing with largely perfect crystals in which the bulk of the diffraction will be dynamical, and since the condition for breakdown of dynamical contrast is a local orientation change, there is little difference in principle between orientation contrast and extinction contrast (Dudley[1]), but the historical terminology persists.

12.2.1 Images of Dislocations

Dislocation images in single-crystal x-ray diffraction images differ from those in electron micrographs primarily because the incident beam cannot be regarded as a plane wave. Of the three types of transmission image characterized by Authier,[2] the most common one seen in wafer defect metrology is the direct image. This appears dark against the perfect crystal background on the topograph (e.g., Figure 12.4). The origin of this enhanced intensity lies in x-rays that are outside the range of diffraction of the perfect crystal. Although not diffracted by the perfect crystal, they will be diffracted by the deformed region around the dislocation. Having suffered no extinction, they are more intense than rays diffracted by the perfect crystal, and the so-called direct image appears as a dark line on the image*.

In thick crystals in transmission, the rays far from the Bragg angle that contribute to the direct image are fully absorbed and the direct image does not appear. The image type is then called dynamical image; defects block the transmission of the beam and dislocations appear as a less intense white line on the image. This can be seen in, for example, transmission images in the heavier compound semiconductors, such as GaAs. In crystals of intermediate absorption, complex dark–light interference patterns can be seen around the line image.

The important direct image and its width are worth discussing more fully. As an approximation, we can consider the crystal to be made up from three distinct regions, namely, the perfect crystal above the defect, the perfect crystal below the defect, and the deformed region around the defect (Figure 12.5a). We set the limit of the deformed region as that where the effective misorientation, (Δ), around a defect exceeds the perfect crystal reflecting range:

$$\delta\left(\Delta\theta\right) = -\left(k \sin 2\theta_B\right)^{-1} \partial\left(\underline{g}\cdot\underline{u}\right)/\partial s_g \tag{12.1}$$

The region around the defect is assumed to diffract as a small mosaic crystal of thickness Δ. The x-rays diffracted from this region are at angles that are not diffracted by the perfect crystal. The integrated intensity in transmission is shown in Figure 12.5b. It varies linearly with thickness for small values of thickness, but as the thickness increases, the gradient decreases and oscillates about a constant value.** As the gradient of the I vs. t curve is everywhere less than that at the origin, the intensity I_t due to material thickness t is always less than the sum of I and I_{t-} from separate regions of thickness t and t–. In Figure 12.5b, where $t/\xi_g \approx 3$, corresponding to the third minimum, it is very obvious that

$$I_t < I_\Delta + I_{t-\Delta} \tag{12.2}$$

There is thus always enhanced intensity around the defect, the contrast being a maximum when the thickness correponds to the first (Pendellösung) minimum, $t/\xi_g \approx 0.88$. For the reflection case and zero absorption (Figure 12.5c) the intensity is independent of thickness beyond a crystal thickness of about an extinction distance ξ_g. Once again we see that for crystal thickness greater than this value, $I_t < I + I_{t-\Delta}$, Equation 12.2 is valid and the defect shows enhanced intensity. In crystals less than an extinction distance in thickness,

* It is conventional in x-ray imaging to display images so that higher intensity corresponds to a darker region. This is the opposite convention from transmission electron microscopy.

** When absorption is included, there is superimposed a gradual decay of intensity with thickness.

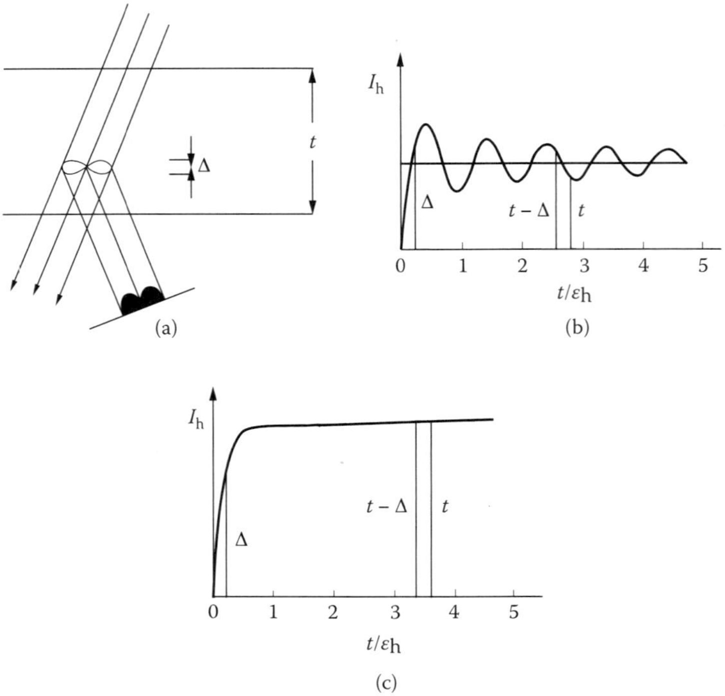

FIGURE 12.5
(a) Diagram showing the splitting of the crystal into perfect material above and below the defect and distorted material close to it. (b) Integrated intensity, Laue case. (c) Integrated intensity, Bragg case.

Equation 12.2 becomes an equality, and defect images are not observed. This has important implications for studies of very thin films on semiconductors that are not at least semicoherent with the substrate, such as thin SOI layers. The danger is that crystals can appear perfect when in reality they may contain defects. Note that dislocations in epitaxial films are never missed from this cause, since the strain field of the film is coherent with that of the substrate and the image is formed from the ensemble of the two.

The width of the image can be deduced using this simple idea of contrast being formed when the misorientation around the defect exceeds the perfect crystal reflecting range, $\delta\omega$. We consider the case of a screw dislocation running normal to the Bragg planes, where the line direction vector **l** coincides with the diffraction vector **g**. The effective misorientation, at distance r from the core is

$$\delta\theta = b/2\pi r \tag{12.3}$$

where b is the magnitude of the Burgers vector **b** of the dislocation. The width of the dislocation image D, which is twice the value of r for which $\delta\theta = \delta\omega$, is thus

$$D = b/(\pi\delta\omega) \tag{12.4}$$

As for a symmetric reflection, $\delta\omega = 2/\xi_g$, we have $D = gb\xi_g/2\pi$. This result can be generalized to

$$D \approx \mathbf{g{\cdot}b}\ \xi_g/2\pi \quad \text{for screw dislocations} \tag{12.5}$$

$$D \approx \mathbf{g{\cdot}b}\ \xi_g/\pi \quad \text{for edge dislocations}$$

As the extinction distance is on the scale of micrometers, so dislocation image widths are of that order. In terms of the structure factor F_g we have

$$D \text{ constant } [\lambda F_g]^{-1} \tag{12.6}$$

The dislocation image width goes down for increasing wavelength and strength of the reflection. Equation 12.5 suggests that the image width is zero when $\mathbf{g{\cdot}b} = 0$. This is of course the classic criterion, originally applied to transmission electron microscopy, where the effective misorientation of the distortion around the dislocation is zero. As the Bragg planes are not distorted or tilted, the dislocation is invisible in that reflection. Strictly, this criterion is that both $\mathbf{g{\cdot}b} = 0$ and $\mathbf{g{\cdot}b} \times \mathbf{l} = 0$, except when the dislocation runs parallel to a high-symmetry axis, and it is valid only for isotropic elasticity. Nevertheless, the contrast is often weak for just $\mathbf{g{\cdot}b} = 0$, and this enables $\mathbf{b}$ to be determined on purely geometric grounds by finding two reflections in which the dislocation is invisible or very weak. This is not normally required in production monitoring, but may be relevant to process development.

This discussion shows that the higher the intrinsic sensitivity to strain, the wider are the images of defects. In practice, the dislocation image widths in semiconductors are of the order 1 to 20 μm. This is orders of magnitude wider than those in TEM images and has three consequences:

- It is relatively easy to image and to map low densities of dislocations, even single dislocations, with low-magnification images.
- It is impossible to resolve individual dislocations where their density is above about 10^4 cm^{-2} in wafers.
- Dislocation bundles are not resolved into individual dislocations but can be imaged as a bundle by the change in intensity.

Use of glancing incidence has a beneficial effect on the image contrast, as well as having the effect of expanding the beam area. If the reference crystal has a grazing incidence beam, the exit beam divergence is reduced by $\sqrt{b}$, the square root of the asymmetry factor. The volume of defective material around individual defects is therefore enhanced over the symmetric geometry. In grazing incidence on the specimen, both the absorption depth and

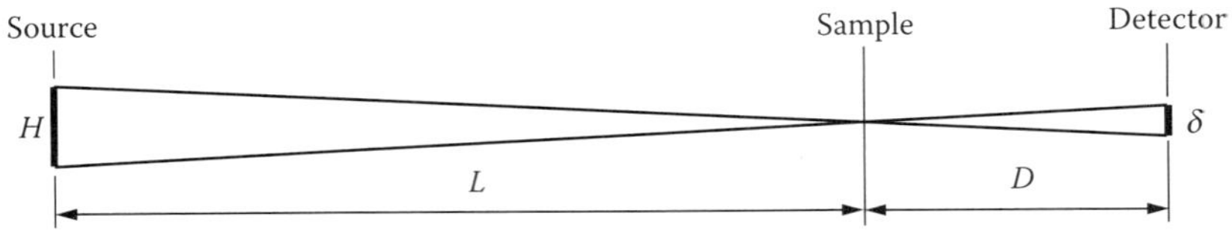

FIGURE 12.6
Schematic diagram showing the geometrical resolution limit set by the projected source height normal to the incidence plane.

the extinction length become small. Thus, both within and away from the Bragg reflecting range the x-ray wave does not penetrate deep into the crystal. The depth of defective material around the defect becomes comparable with the depth of penetration of the x-ray wave, and almost 100% contrast can be obtained from the defect. This is especially relevant to images from thin isolated films.

The contrast of defects in x-ray topographs has been recently reviewed by one of the authors[3] to which the reader is referred for further details and a bibliography.

12.3 Spatial Resolution in XRDI

We have seen that dislocation images, usually the narrowest type of defect image, are some micrometers across. We need sufficient resolution to observe and resolve these images, but there is little point in having resolutions below about 1 μm. The next issue is, What spatial resolution is achievable? The limitations are the projected size of the x-ray source and the detector resolution.

To a good approximation, the spatial resolution r normal to the incidence plane is related to the specimen-to-source distance, D, the specimen-to-detector distance, L, and projected height of the source, H, by the simple geometrical relation (Figure 12.6)

$$r = H\,L/D \tag{12.7}$$

The benefits of a microfocus source, and of being able to place the detector very close to the specimen, are immediately clear. For example, with a projected source height of 0.04 mm, distance of 160 mm from the specimen, a spatial resolution of 1.25 μm can be achieved for a 5-mm specimen-to-detector distance. In practice, specimen–detector distances are significantly larger than this, with a microfocus source simply to avoid mechanical clashes with a large wafer, and the source contribution to resolution loss is negligible. However, using larger sources such as high-power rotating anodes, the source might be as much as 3 m away from the specimen to obtain comparable resolution.

With a microfocus source the dominant limitation on spatial resolution is that of the detector. X-ray magnifiers are just possible, but are neither efficient nor convenient and reliable. We therefore are limited to the direct sensor resolution. The best resolution is still nuclear photographic plates, at approximately 0.5 μm, but these require several hours of wet processing and have high consumable cost. Fiber optically charge coupled devices (CCDs) have now achieved 3-μm pixel resolution (Section 12.3.1), are second only to nuclear plates or lithographic films, and are convenient and practical to use in the fab. They may also be constructed with large areas at lower resolution to image whole wafers in a scanning method.

12.3.1 Real-Time Image Detectors

Real-time x-ray imaging detectors now almost all use an x-ray-to-optical converter, followed by a CCD sensor, as illustrated in Figure 12.7. For the very highest (submicron) resolution, the fiber-optic coupling is replaced by a lens coupling, but this leads to a large loss of efficiency and is only feasible at synchrotron radiation installations. Commercial detectors are available at the time of writing with a 3-μm pixel resolution. Below this figure, cross talk between fibers worsens the point-spread function and limits the resolution.

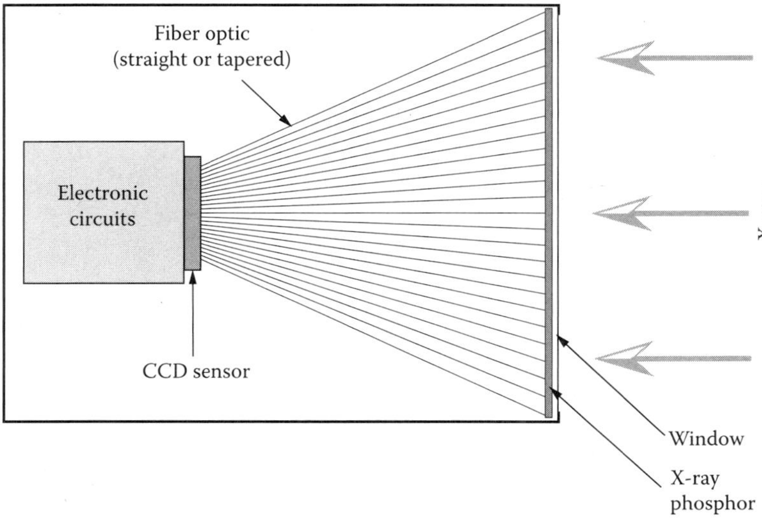

FIGURE 12.7
Schematic construction of an x-ray imaging detector.

There is always a conflict between the requirements of high spatial resolution, large field of view, and high efficiency, which is separate from the technological issue of what CCD sensors are currently available. In order to see why this conflict arises, we consider a flux I/unit area/unit time of x-ray photons incident on a detector of efficiency η. The number of photons N detected in an integration time τ in a square picture element of side ε is then

$$N = I\eta\varepsilon^2\tau \tag{12.8}$$

If two elements in the topograph have an intensity difference ΔI, then the difference in signal ΔN is

$$\Delta N = \Delta I\eta\varepsilon^2\tau \tag{12.9}$$

When written in terms of the contrast C, defined as $C = \Delta I/(2I + \Delta I)$, this becomes

$$\Delta N = \eta\varepsilon^2\tau 2IC/\left(1 - C\right) \tag{12.10}$$

Now the root mean squared (rms) noise on the signal is $(2N + \Delta N)^{1/2}$, since it is a Poisson distribution, and the signal/noise R is then given by

$$R = \Delta N\left(2N + \Delta N\right)^{-\frac{1}{2}} = \left(C\Delta N\right)^{\frac{1}{2}} \tag{12.11}$$

Thus,

$$\varepsilon = \frac{R}{C}\left(\frac{1 - C}{2\eta\tau I}\right)^{\frac{1}{2}} \tag{12.12}$$

This is a form of the Rose-de-Vries equation, and we see that it sets a fundamental limit on the spatial resolution of a quantum-limited system. The spatial resolution is improved by increase of the incident intensity I, the detector efficiency η, and the integration time τ. To retain the high spatial resolution, the scintillating phosphor must be thin; to increase efficiency, it must be thick. Progress in detectors in recent years has involved the development of heavy, fine-grained, or single-crystal rare-earth-based phosphors, and the development of photon-counting CCD sensors.

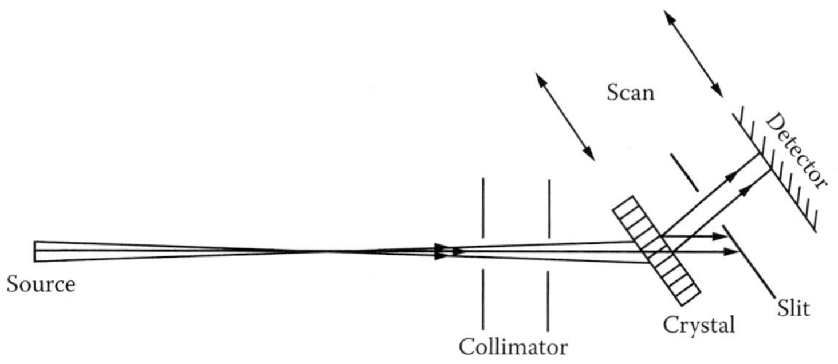

FIGURE 12.8
Schematic diagram of the projection topography configurations. For projection topographs, the specimen and film are translated across the beam. (Note that very narrow (ca. 10 μm) slits are used for section topography, whereas those for projection topography are typically 0.25 mm.)

12.4 X-ray Defect Imaging Methods

12.4.1 Lang Projection Topography

The foundation for x-ray defect mapping methods is a scanning technique devised by Lang.[4] In order to obtain an image of the whole crystal and still retain high spatial resolution, Lang devised a goniometer in which the crystal and film were translated together across the beam (Figure 12.8). Lang cameras, which are two-circle goniometers equipped with a precision translation stage and adjustable (or interchangeable) incident beam slits, are available commercially, and the technique has become the most widely used laboratory technique of x-ray diffraction imaging. A very large body of experimental data and theoretical studies for contrast analysis is available for this method.[2] It is capable of production of excellent, detailed images of crystals, in which individual dislocations are visible at a level of detail down to approximately 1 μm.

Several problems arise if this method is considered for semiconductor fabrication lines:

- The mechanics of scanning a 300-mm wafer do not permit a close approach of the film to the specimen. As we have seen in Equation 12.7, for good resolution this will imply a large source–specimen distance, and hence a tool with a large footprint (several meters long).
- Film is an unacceptable medium for recording data in a fab. Direct digital recording with information extraction is required. Though image plates can be used, their resolution is not good enough for all

defect mapping work. High-resolution CCDs are not available, since the detector has to be the same size as the wafer in this method.

- Overall cycle time is poor.
- The Lang geometry is well suited to transmission, but not to reflection imaging. The latter is particularly important for epitaxial films and devices.

These considerations stimulated the development of the fully digital systems described in the next section.

12.4.2 The BedeScan Method

This method was devised by Bowen, Wormington, Feichtinger, and Pina[5,6] of Bede plc. It is illustrated in Figure 12.9. The x-ray beam is diffracted from the wafer in either reflection or transmission, and the diffracted beam recorded on a two-dimensional position-sensitive (imaging) detector.

In reflection using CuK radiation, the image at the detector is a double line of $K\alpha_1$ and $K\alpha_2$, as shown in Figure 12.10, with each line imaging a different part of the wafer. A bright dot is seen on the $K\alpha_1$ line in Figure 12.10, which is the image of the cross section of a defect; this is not seen on the $K\alpha_2$ line, which images a different area. In transmission, using Mo radiation, the image at the detector contains the $K\alpha_{1/2}$ (normally unresolved) and $K\beta$ lines.

In the Lang method, a slit is used to eliminate $K\alpha_2$, since it otherwise causes blurring of the image. Much shorter distances from specimen to detector can

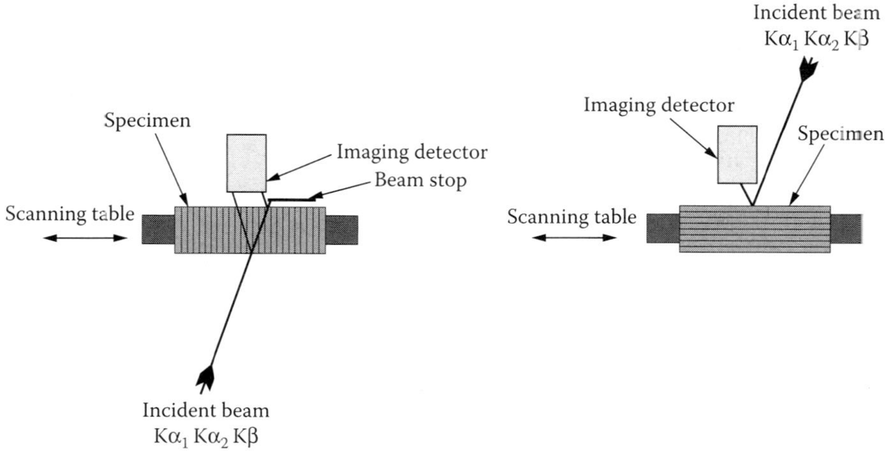

FIGURE 12.9
The new scanning XRDI method in (a) transmission and (b) reflection. The imaging detector is fixed, and the wafer (specimen) scans through the beam.

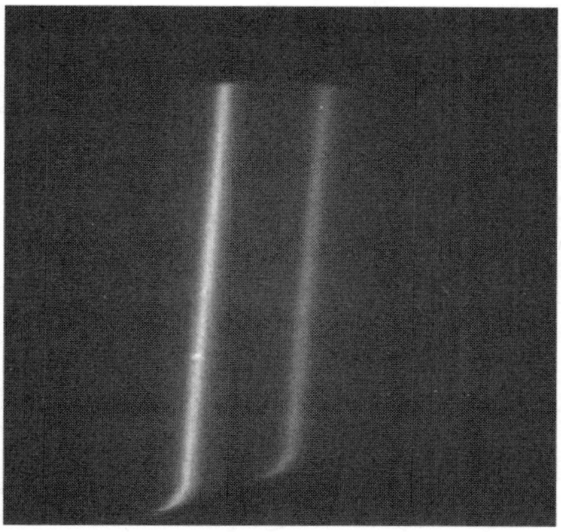

FIGURE 12.10
Single-frame" image showing Kα1 (left) and Kβ (right) lines. These two lines image slightly different parts of the wafer.

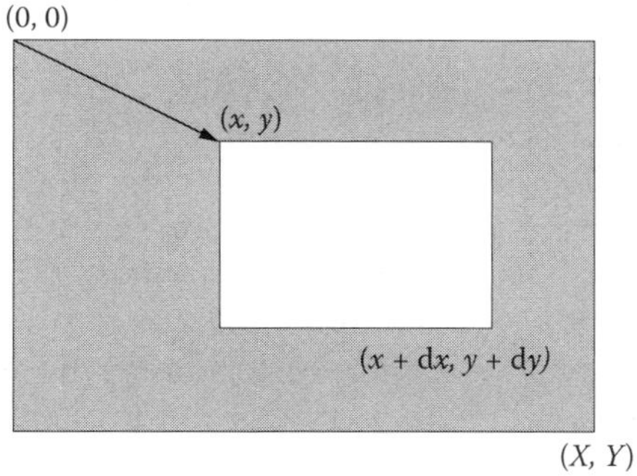

FIGURE 12.11
Illustration of principle of virtual scan of detector.

be used in the new method, since there are no diffracted beam slits to interfere, and the blurring is lower. We can also completely eliminate the spectral blurring in reasonably good crystals, as will be shown below. Therefore, significantly more intensity is available by including the whole of the Kα_1 and Kα_2 lines.

When the wafer is scanned by a mechanical step, the new image is superimposed on the previous one. In order to integrate each frame into a complete image, we therefore undertake a virtual scan of the detector in the computer. This is illustrated in Figure 12.11. The area (x, y) to $(x + dx, y + dy)$ is the image acquired in single frame. The area (0, 0) to (X, Y) is the full image of the wafer, held in the computer. The frame currently received from the detector is mapped into the correct part of computer memory, so that scanning across the whole wafer builds up the correct image of the wafer. This principle can in fact be applied to any kind of image (e.g., an optical micrograph or

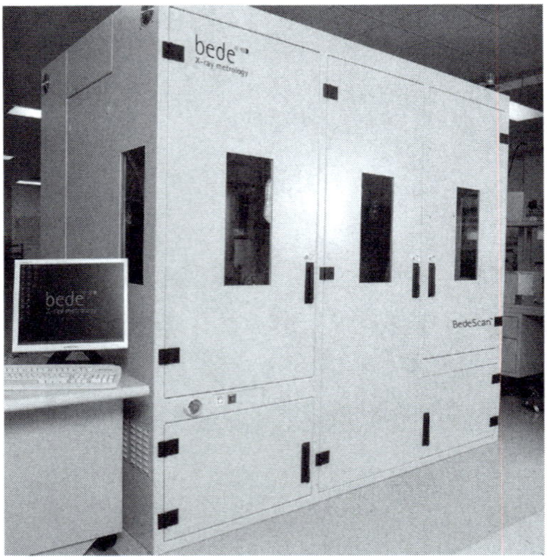

FIGURE 12.12
BedeScan instrument in fab research lab configuration.

x-ray radiograph) and allows a limited-area detector to be used to image a specimen of unlimited size at high resolution. In addition to the image contrast information, the absolute position of the Kα lines on the detector are a sensitive measure of the local orientation of the wafer, specifically its deviation about an axis normal to the incidence plane. These orientation data may also be conveniently stored and mapped to give a quantitative map of specimen distortions, e.g., bowing or distortion in processing.

Figure 12.12 shows a BedeScan instrument. X-rays are produced using an 80-W microfocus source. In reflection, the incidence angle (θ) of the CuK x-ray beam on the wafer is adjusted to satisfy the Bragg condition by rotating the source, which is carried on a compound XY mechanical slide above the surface of the wafer. In the transmission geometry, the MoK source is below the surface of the wafer and a beam stop is placed between the wafer and the active area of the detector to occlude the direct x-ray beam. In both geometries, the CCD detector is mounted above the wafer, also on a compound axis.

The wafer is horizontally mounted using a three-point kinematic clamp. The wafer clamp is carried on two high-precision translation (y and x) axes and a rotational (ϕ) axis, ordered from bottom to top of the instrument. The wafer is aligned by rotating the ϕ axis until the diffraction stripes (as shown in Figure 12.10) are parallel to the y axis. Under computer control, the wafer is scanned through the incident x-ray beam using the x axis and stepped using the y axis in order to collect a partial topograph (stripe) with a height approximately equally to the height of the active area of the CCD detector. These stripes are then tiled together to create the final x-ray topograph. The alignment of the system rotating the specimen until the diffraction stripe (Figure 12.10) is vertical ensures that the stripes may be stitched together accurately without distortion.

One important point about this method is that the single-frame image is available for image processing before it is integrated into the stripe. This is not possible with photographic or image plate methods in the Lang camera. The kind of processing that can be performed is as follows:

- Correction for spectral spread. The stripes shown in Figure 12.10 are each a true image of the sample but at slightly different wavelengths. Hence, the diffracted angles are slightly different. In film integration, this appears as a double image; thus, the $K\alpha_2$ is normally filtered out with a slit. In the BedeScan method, the image can be altered in magnification in the *X* direction only before integration. This compensates for the spectral blurring and avoids image doubling and loss of resolution.
- Correction for distortion. The projection effect of the angle of the diffracted beam can be calculated and eliminated, so that images are true to shape.
- Virtual slit to control orientation contrast.[6] The software must select the region of interest (ROI) on the frame (Figure 12.10) before integration. If the orientation of the specimen changes slightly between frames, then the stripe moves. A narrow ROI will therefore give greater sensitivity to orientation contrast. This virtual slit method can be used to postprocess the same data with different degrees of orientation contrast. A moving or wide ROI allows automatic compensation for the imaging of warped wafers (this requires mechanical Bragg angle control in conventional topography).

The imaging properties of the resulting wafer defect images are discussed in the references, and in summary are:

- Image contrast shows both extinction and orientation contrast, in a manner very similar to that in white beam synchrotron radiation topography,[8] but with control over orientation contrast through the virtual slit concept.
- The resolution is limited only by the detector, down to a resolution of <0.5 μm.
- Exposure time is shorter than on a conventional film system. A 300-mm wafer can be scanned for slip lines and other dislocation bundles in approximately 30 min.

12.4.3 Section Topography

This transmission method[2] provides an image of a section through the crystal and as such enables the experimenter to study the three-dimensional distribution of defects. The beam from the spot of a fine- or micro-focus source is collimated into a ribbon beam of width approximately 10 μm before the single-crystal specimen. This provides an incident beam of width that is small compared to that of the base of the Borrmann fan formed by extremes of the diffracted and transmitted beams with the crystal surface. (In other words, the beam width must be much less than $t \sin 2\theta_B$, where t is the specimen thickness and θ_B the Bragg angle.) The specimen is adjusted until a strong diffracted beam from the characteristic $K\alpha_1$ line is obtained for the diffraction planes chosen and the photographic plate or high-resolution image detector placed behind the specimen. A diffracted beam slit prevents the main beam from striking the detector.

12.5 Example Applications

A defect map of a whole 200-mm wafer is shown in Figure 12.13. The points that can be observed on this image are:

- Growth dislocations cluster in the center of the wafer.
- Thermal slip lines (dark lines consisting of dislocation bundles) are generated from bulk micro-defects (BMDs) (oxygen precipitates) on orthogonal slip planes after thermal treatment.
- The notch is a strong generator of slip lines.
- Slip lines sometimes extend all the way to the center of the wafer.
- At higher magnification, the detailed dislocation structures can be seen (Figure 12.4).

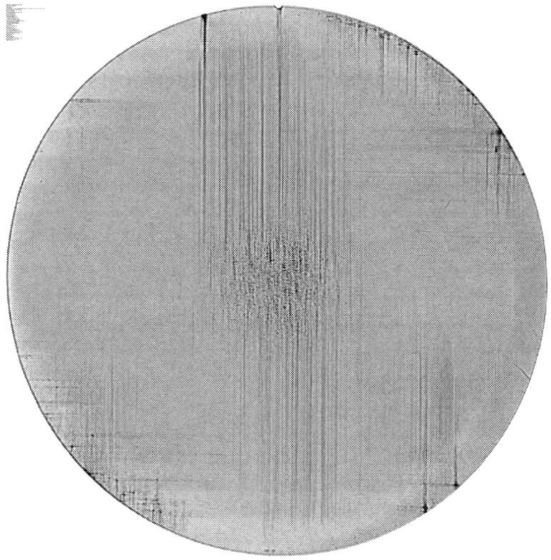

FIGURE 12.13
Whole-wafer defect map. 200 nm wafer, BedeScan image, MoKα 220 reflection in transmission.

The importance of (avoiding) slip lines in semiconductor wafers is that they attract impurities, which change the local band structure and electron mobility. They can effectively short-circuit transistors that they cross. In compound semiconductors, they are also nonradiative recombination centers, which decrease the efficiency of optoelectronic devices.

Defect recognition can be automated quite simply. We may write

$$I_{total} = I_p + nI_d \tag{12.13}$$

That is, the total integrated intensity I_{total} in a region is the sum of perfect crystal scattering, I_p, plus (incoherent) dislocation scattering. The latter is approximately linear with the number of dislocations, n, in the area illuminated and I_d, the scattering per dislocation. This is because (in the reflection or low absorption transmission cases used here) each dislocation scatters a small amount of energy from the parts of the beam that are not diffracted by the perfect crystal. Even though the visual images may superimpose within a dislocation bundle, the dislocations scatter independently and to a good approximation obey the kinematic theory (Chapter 11), up to the point at which their strain fields overlap. Dislocation images map the strain field up to a maximum of about 20 μm from the dislocation core,[9] so we may estimate that overlap will occur at a local dislocation density of approximately 10^5 cm^{-2}. Thus, the excess intensity above the perfect crystal scatter is a quantitative measure of dislocation density, if these are substantially the only defects present. Figure 12.14 shows this concept applied to the whole-

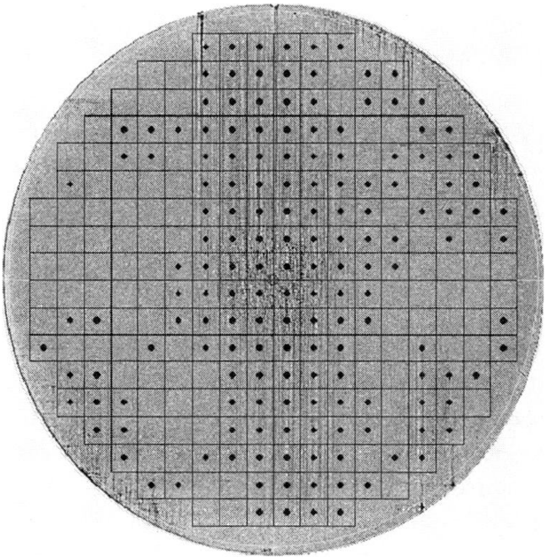

FIGURE 12.14
Automatic defect recognition. The image of Figure 12.13 with a superimposed grid corresponding to a microchip die layout. Black circles are automatically superimposed on dies in which the integrated intensity (corresponding to dislocation and other defect density) is over a threshold level.

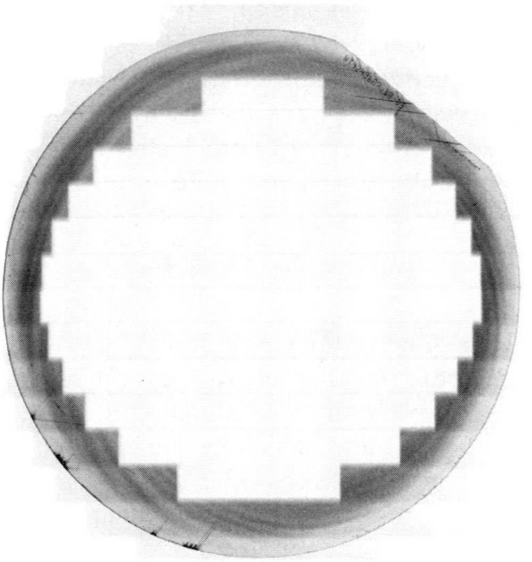

FIGURE 12.15
Edge scan showing defects arising from edge damage. BedeScan image, MoKα 220 reflection in transmission. (Courtesy Dr. D.-L. Lee, LG Siltron.)

wafer sample. A grid corresponding to the chip die size is superimposed on the image, and the dies in which the integrated intensity falls above a certain threshold are marked with a black dot. It is known from studies over many years, for example, Matthews and Blakeslee[10] and Feichtinger,[11] that most dislocations cause either a direct loss of yield or a loss of performance, e.g., through reduction in carrier mobility, or at least a reduced chip service life. This map is therefore a direct predictor of one aspect of manufacturing line yield. The metric will saturate as dislocation density increases to about 10^5 cm^{-2}, but this is likely to be well above the level at which devices are affected.

Since by far the most slip lines arise at the edge of the wafer, very probably from grinding and polishing damage that has not been sufficiently etched away, it makes sense to restrict a scan to the edge region. This increases the throughput substantially. An example is shown in Figure 12.15. Manufacturers have rigorous criteria for the yield at the edge of a wafer, and this image shows clearly why the problem exists. Only about 20% of dies made in the region imaged would be expected to be successful. Even apparently innocuous features such as the laser-etched identification mark, and the ground flat or notch, are seen to be capable of slip line generation after some thermal treatment.

Relaxation after epitaxial deposition is another generator of dislocations. Arrays of 60° mixed dislocations generated by epitaxial mismatch in a Si-p/Si-p^{++} system epilayer are shown in Figure 12.16. The nonuniformity of their distribution is because they have been nucleated at points of mechanical damage at the edge of the wafer. A section topograph of this

FIGURE 12.16
Epitaxial misfit dislocation network in a Si substrate/Si epilayer system, taken in reflection, with the 115 reflection. Sample is p/p++ Si (i.e., with a heavily B doped substrate), showing 60° mixed misfit dislocations at the interface with the 6-μm-thick epilayer. (Courtesy Wacker Siltronic Corp., OR.)

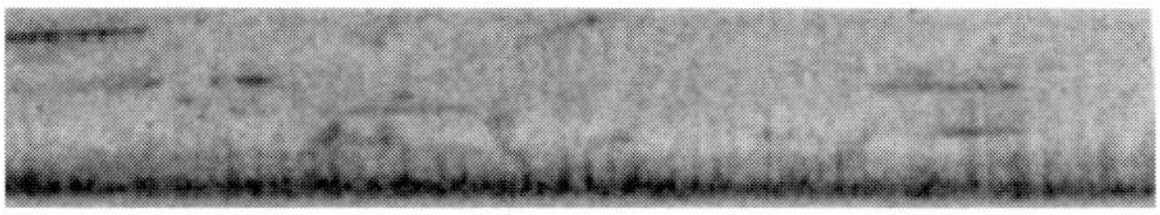

FIGURE 12.17
Section topograph (MoKα 220) of a sample with heavy thermal slip, showing the distribution of defects from top to bottom across a slice through the wafer. The height of the image is the thickness of the wafer.

sample is shown in Figure 12.17, in which the distribution of defects from front to back can be seen.

Sometimes the relaxation is introduced deliberately, to form a pseudosubstrate for strained silicon deposition on a wafer scale. In this case it is valuable to be able to check the uniformity of the relaxation. While this can be measured by the HRXRD methods discussed in Chapter 4, scanning the complete wafer in a BedeScan tool is far faster, and is done to a resolution finer than a scribe line.

We can use defect mapping even in cases where the epitaxy is performed only in a selected area. Figure 12.18 shows 300-μm squares in which SiGe has been deposited. The crosshatch network of dislocations caused by relaxation is easily seen. The integrated intensity from this region can also be measured and used as a quality control.

Finally, we are not, of course, restricted to silicon. Figure 12.19 shows a wafer of GaN on SiC, important in the blue laser and high-temperature semiconductor device markets. This is much less perfect than silicon substrates, and the defects are very clearly mapped by the XRDI method. They include dislocation bundles that form micropipes and also subgrain boundaries.

FIGURE 12.18
Small area scan of SiGe selected-area epitaxy at 3-μm resolution. The crosshatched pattern in the 300 × 300 μm epitaxial areas is an image of interface dislocations, showing that the SiGe is slightly relaxed. The film thickness and the relaxation decrease from (a) to (c) in the three images shown. CuKα, 004 in reflection. (Samples courtesy Wiebe de Boer, Philips Semiconductor, Fishkill, NY.)

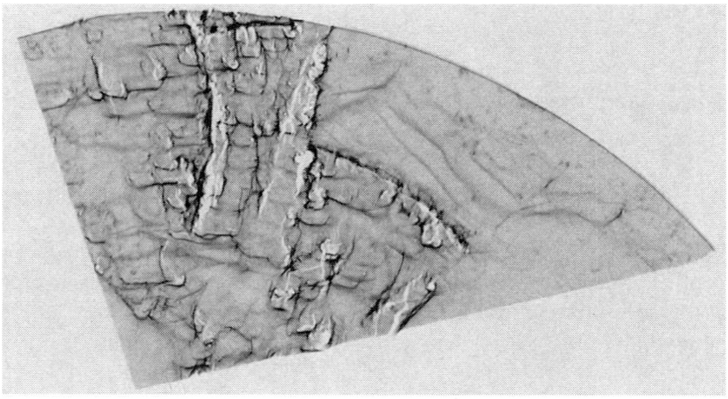

FIGURE 12.19
Image of SiC substrate wafer with a GaN epilayer, with the diffraction angle set on SiC.

12.6 Summary

- XRDI with digital imaging is now capable of use in fab research and production facilities.
- XRDI reveals dislocations, whether in slip lines or isolated individual areas, in all cases where the line length is more than a few micrometers.
- A particularly useful application to low dislocation densities is the detection of the onset of relaxation, which cannot be done with the usual HRXRD methods.

- XRDI reveals precipitates such as are produced in oxygen-rich regions in silicon.
- XRDI also shows surface stress concentration points such as scratches and handling/supporting zones.
- XRDI also reveals grinding and polishing damage, especially at edges, notches, and wafer identification marks.
- A whole-wafer scan suitable for revealing slip lines takes approximately 30 min for a 300-mm wafer.

References

1. M. Dudley, X.R. Huang, and W. Huang, *J. Phys. D: Appl. Phys.*, 32 (1999) A139–A144.
2. A. Authier, *Dynamical Theory of X-Ray Diffraction*, Oxford University Press, Oxford, 2001.
3. B.K. Tanner, in *X-Ray and Neutron Dynamical Diffraction: Theory and Applications*, A. Authier, S. Lagomarsino, and B.K. Tanner, Eds., Plenum Press, New York, 1996.
4. A.R. Lang, *Acta Cryst.*, 12 (1959) 249.
5. D.K. Bowen, M. Wormington, and P. Feichtinger, *J. Phys. D, Appl. Phys.*, 36 (2003) A17–A23.
6. D.K. Bowen, M. Wormington, L. Pina, and P. Feichtinger, X-Ray Topographic System, U.S. Patent 6,782,076, granted 2004.
7. M. Wormington, P. Feichtinger, and D.K. Bowen, Virtual X-Ray Slits, provisional U.S. Patent submission.
8. B.K. Tanner and D.K. Bowen, *Mater. Sci. Rep.*, 8 (1992) 369–407.
9. J.E.A. Miltat and D.K. Bowen, *J. Appl. Crystallogr.*, 8 (1975) 657–669.
10. J.W. Matthews and A.E. Blakeslee, *J. Crystal Growth*, 27 (1974) 118–125.
11. P. Feichtinger, Ph.D. thesis, University of California, Los Angeles, 2000.

Part 3

The Technology

13

Modeling and Analysis

In order for any metrological method to be usable in the fab, there must be an automated and reliable method of extracting the data without any operator intervention or (worse) personal judgment. As with many methods of optical metrology, x-ray metrology (XRM) is not a ruler, nor is it a system with an output that is monotonically related to thickness or any other parameter. Interpretation from x-ray scattering theory is required. This chapter discusses the methods by which this may be achieved. Fortunately, as has been discussed in Part 2 of this book, x-ray theory is extremely sound and comprehensive. It is based upon few assumptions, all of which are easily validated in practice. In particular, the x-ray scattering factors are well known. As a consequence of the high energies of x-rays in comparison with energies of valence effects in solids, they are invariant with chemical or physical state to sufficient accuracy. We can therefore assert with confidence that if we have a wafer whose film structure is known, and a metrology tool whose characteristics, such as beam size and divergence, are known, we may calculate the expected diffraction or reflectivity pattern with high accuracy.

13.1 What Has Been Measured?

We have measured the x-ray scattering power of a sample, usually as a function of angle of incidence or scatter, or both. We have seen that such measurements are sufficient to lead to the metrological sensitivity, precision, and accuracy that are needed. Unfortunately, we cannot transform the data directly into the structure and its dimensions. This is for two reasons:

1. We have only measured the intensity of this scatter and not its amplitude and phase. Both of these would be required in order to make a direct reconstruction of the structure of the material from the scatter.
2. We have only measured the scatter around a strong diffraction or reflectivity peak, and not over a 2π solid angle.

We need not, however, generally be concerned about the uniqueness of our solution. This is because we do not start from complete ignorance of the structure, but from a very close knowledge of what the manufacturer was trying to make. It is normally possible to state whether or not this has been achieved. And with the realistic constraints of the manufacturing process, ambiguity in the final result is rarely a significant issue, so long as the data are sensitive to the parameters of interest.

The methods of analysis of the x-ray data are therefore based upon two general methods:

1. A partial application of the direct method approach, extracting periodicities from the data that correspond to layers, or combinations of layers, in the film stack
2. Simulation of the x-ray scattering from a model structure, followed by comparison with the data and refinement of the model

The mathematical techniques applied were originally developed in other branches of applied science, though the most successful of the modeling methods, differential evolution (a type of genetic algorithm), had one of its earliest applications to the case of x-ray scattering.

13.2 Direct Methods

If we glance at an x-ray diffraction (XRD) or x-ray reflectivity (XRR) pattern containing interference fringes, the periodicities seem obvious. Newcomers to the field often assume that a simple Fourier analysis will suffice to pull out the periodicities automatically. However, the problem is not so simple, as has been shown by Hudson et al.[1] There are several problems:

1. The interference fringes are quite weak, and relatively few are observed. The intensity of the Fourier peak is therefore low and may be obscured by noise (a low-noise detector is essential).
2. The transform is dominated by the sharp substrate peak (in diffraction) or the $2\theta^{-4}$ decay (in reflectivity), which scatter intensity throughout Fourier space and tend to mask any other modulation.
3. The whole data set is of limited extent, so truncation peaks arise in the transform.
4. In XRR the fringes are not perfectly periodic but bunch up toward the low-angle side. This can be corrected by measuring the critical angle and applying refractive index correction.[2,3] However, this is

not a sharp transition, and its measurement is uncertain except by modeling the whole curve.

The kind of procedures that show success are:

1. Preprocess the data. In diffraction, mask out the substrate peak from the rocking curve, joining smoothly across the gap; in XRR, determine the average decay curve (which is only $(2\theta)^{-4}$ for a perfectly sharp surface) and subtract this from the data.
2. From the average values of the curve determine a background to the interference oscillations, and divide the data by the smoothed curve.
3. Subtract the residual background.
4. Perform an autocorrelation on the processed curve, to reduce the noise.
5. Perform the Fourier transform.

With this means, periodicities due to the primary film thicknesses in the data are usually extracted. However, some of the peaks that appear in the transform are difference or beat frequencies, or those due to a film stack, and do not necessarily correspond to a single layer. For example, the Fourier peak from a Ti/TiN barrier layer measured in XRR will consist of a strong peak due to the sum of the two layers, and very weak peaks due to the individual layers, because of the low contrast between their electron densities. Moreover, careful preprocessing of the data is usually required, to avoid the strong θ^{-4} decay in XRR and the substrate peak in XRD from dominating the transform.

The conclusion is that Fourier analysis, with the above procedure, is a powerful aid to a skilled researcher. It can be applied to automated analysis in simple cases, in which the analysis may need to be set up *ad hoc* for the film stack concerned. However, it is not appropriate for automated analysis in more complex or general cases. Its main advantage over modeling methods is very high speed computation.

More complex integral transforms have been studied, e.g., the two-dimensional Wigner transform,[4] but so far without success. Even the best-quality data are too sparse and noisy to give reliable information when transformed in this way. The most promising of the transform methods appears to be that of wavelet analysis.[5] This is an extension of the Fourier transform method in which each layer is represented by a limited-range wavelet corresponding to its length scale and roughness. This appears to avoid the worst difficulties of the Fourier transform and has been valuable in obtaining at least good trial parameters for further refinement.

13.3 Data-Fitting Methods

After studying both direct and data-fitting methods, as well as expert systems, over three decades, our opinion is that modeling methods, especially those based upon genetic and evolutionary algorithms, are the most reliable and repeatable way to extract metrological information from x-ray data. The following treatment is based largely on the work of Wormington, Panaccione, Matney, and Bowen.[6–8]

Automated fitting of a model to a data set is of course a common problem in science and engineering. The field of data fitting and parameter optimization has a long and fruitful history. The earliest successes were for linear problems that possessed a single minimum in the objective function *f*, which represents the difference between the simulation and the data. The mean squared difference (MSD) between the experimental and simulated data was commonly used as the objective function because of its computational simplicity in the days before fast digital computers. Unfortunately, the MSD has three problems:

1. A rounded minimum, which slows the final stage of convergence
2. Excessive emphasis on outliers in the data (which always occur with Poisson noise)
3. Poor handling of data that have a very wide dynamic range

More recent research has focused on nonlinear problems and on those with local minima in the objective function in addition to the global minimum. A variety of data-fitting and parameter optimization strategies have been developed for such systems.[9,10] Those most commonly encountered, and their strengths and weaknesses, are:

- **Direct search**: The parameter space is divided up into small but finite regions. The objective function is calculated for each region, and the region that gives the smallest value for *f* is said to give the best-fit (optimum) parameter values.
    - This is a very reliable method if applicable. However, the parameter space is multidimensional and simply too vast for direct searches in realistic cases. It becomes uncomputable for all but trivial cases.
- **Downhill simplex**: An initial guess at the parameter values is made. The simplex (a geometrical construction) then moves in directions that decrease the value of *f*. The parameters that yield the smallest value of *f* in the neighborhood of the initial guess are said to be the best-fit parameters.

- **Levenberg-Marquardt method**: An initial guess at the parameter values is made by the user. The algorithm then combines linearization and gradient searching of the objective function to minimize *f* in the neighborhood of the initial guess. The parameter values giving the smallest value for *f* are then selected as the best-fit parameters.
    - These two methods work well for nonlinear problems because they are guided by the geometry of the objective function in parameter space. The Levenberg-Marquardt method is significantly faster. However, the initial estimate of the parameter values needs to be very close to the optimized values if local minima are present, as they will become trapped in the first local minimum that they encounter. These two methods are therefore only effective when the parameters are initially contained within the multidimensional "well" of the global minimum. In most practical cases in x-ray scattering, we have found them to be of limited use.
- **Monte Carlo method**: The parameter space is again divided into small regions. Regions are selected at random and the objective function is evaluated. After a certain number of regions have been chosen, or when *f* is smaller than some specified value, the algorithm is stopped. The region with the smallest value for *f* is said to yield the best-fit parameter values.
- **Simulated annealing**: This uses the physical principles governing annealing (i.e., the slow cooling of a liquid so that it forms a crystal) to search the objective function and obtain the best-fit parameters. There is a finite probability in any step that the parameters can move in a direction so as to increase *f*, so the method does escape from local minima, but slowly.
    - The Monte Carlo and simulated annealing methods do not get trapped in local minima, and will eventually find the global minimum. However, they are very inefficient at searching the parameter space, since they search it randomly without taking into account the geometry of the objective function.
- **Differential evolution**: A form of evolutionary algorithm. A multidimensional vector represents the parameters to be optimized (thickness, composition, etc.). Two "parent" sets of these vectors are initially constructed by random generation within the bounds of the parameters specified by the metrological recipe. These sets, typically 20 to 100 members, can further alter by random mutation. Combinations of two members of the sets as parents provide slower changes known as inheritance. The current best fit is the vector with the smallest value of the objective function, but this is discarded immediately when a better fit is obtained.

- This method escapes efficiently from local minima and finds the global minimum in a reasonable time. The key to its success is a combination of random search and a framework of guidance by the geometrical shape of the objective function.

This method has been highly successful for interpretation of x-ray data.[11] We shall discuss it in more detail in the rest of this chapter. Every comparison between experimental and theoretical data shown in this book was obtained by means of commercial software using differential evolution to optimize the parameters.

13.4 The Differential Evolution Method

We have employed an evolutionary algorithm (EA) called differential evolution.[12] Through simple mutation, recombination, and selection schemes, parameter vectors with better fitness are found. Fitness is measured through an objective function, f, which expresses in some mathematical form the difference between the simulation and the data set. Mutation is an operation that makes small random changes to one or more of the population vectors. Mutation is critical for maintaining diversity in the population of parameter vectors. Recombination is a complementary operation that creates parameter vectors (offspring) by combining two parameter vectors from the previous generation (parents) and helps focus the search on promising regions of the parameter space. Selection guarantees that the fittest parameter vectors will propagate in future generations. EAs differ from the conventional parameter optimization methods listed above in several important ways:

1. EAs optimize the trade-off between exploring new points in the parameter space (mutation) and exploiting the information discovered thus far (recombination).
2. EAs operate on many solutions simultaneously (implicit parallelism), gathering information from current search points to direct the search. Their ability to maintain multiple solutions concurrently makes EAs less susceptible to the problems associated with local minima and noise.
3. EAs are randomized algorithms in that they use operators whose results are governed by probability, but they do not perform purely random searches (in contrast to the Monte Carlo algorithms).

Let the experimental data contain N measured points (θ_j, I_j), where θ_j is the incidence angle, I_j is the intensity measured at θ_j and $j = 1, 2, \ldots, N$. Simulated data $I(\theta_j,\mathbf{p})$ are computed assuming a structural model with n

continuous, adjustable parameters represented by the vector $\mathbf{p} = [p_1, p_2 \ldots p_n]$ and are compared to the experimental data using some objective function $f(\mathbf{p})$. Guided by $f(\mathbf{p})$, the DE algorithm attempts to optimize the parameter vector **p** starting with an initial population of randomly generated parameter vectors, by a repeated cycle of mutation, recombination, and selection.

The form of the DE algorithm known as *DE/Best/1/Bin* was used in this work. We initialize the population by assigning the parameter vector $\mathbf{p}_0$, the user's initial guess at the structure, while the remaining m vectors (the population) are initialized by assigning each parameter with a randomly chosen value from within its allowed range. The control parameters of the DE algorithm, namely, the mutation constant and crossover constants, respectively designated F and C_r, must be empirically determined to give fast convergence (seconds or minutes depending on the data, model, and computer speed). The population size is also determined empirically, and a value of about 10 times the number of adjustable parameters is usually appropriate.

A detailed flowchart for the DE algorithm used in this work is shown in Figure 13.1.

Once all of the parameter vectors have been initialized, the objective function for each $\mathbf{p}_i$ is evaluated. The parameter vector with the lowest error is stored in the best-fit vector $\mathbf{b} = [b_1, b_2, \ldots, b_n]$. This vector is used to track the progress of the optimization and is updated whenever an equal or better solution than the best-so-far vector is found.

The crucial idea in DE is its simple scheme for creating new population members. Two randomly selected vectors, $\mathbf{p}_a$ and $\mathbf{p}_b$, are chosen from the current population. The difference vector $(\mathbf{p}_a - \mathbf{p}_b)$ is then used to mutate the best-so-far vector, **b**, according to the relation

$$\mathbf{b'} = \mathbf{b} + F\,(\mathbf{p}_a - \mathbf{p}_b) \tag{13.1}$$

where F denotes the mutation constant. The value of F must be empirically selected by the user to give fast convergence, and in this work we have used $F = 0.7$. As the evolving population vectors converge, the differences between them diminish, and hence the difference vector remains scaled to an appropriate size. With **b**' in hand, a trial vector $\mathbf{t} = [t1, t2, \ldots, tn]$, which competes with the vector $\mathbf{p}_0$, is assembled. Starting with the randomly chosen jth parameter, the trial parameters t_j are consecutively loaded (modulo n) from either **b**' or $\mathbf{p}_0$. A binomial distribution is used to decide which parameters come from **b**' and which come from $\mathbf{p}_0$. A random number chosen from a uniform (0, 1) distribution is compared with a user-selected recombination (crossover) constant C. If the random number is less than or equal to C, then t_j is loaded with the jth parameter from **b**'. If the random number is greater than C, then the jth parameter of **t** is loaded from $\mathbf{p}_0$. We normally use $C = 0.5$. After $n - 1$ trials, **t** gets its final parameter from **b**', so that at least one parameter of **t** is different from $\mathbf{p}_0$. With the vector **t** assembled, any constraints are then taken into account. If the value of the trial parameter t_j falls

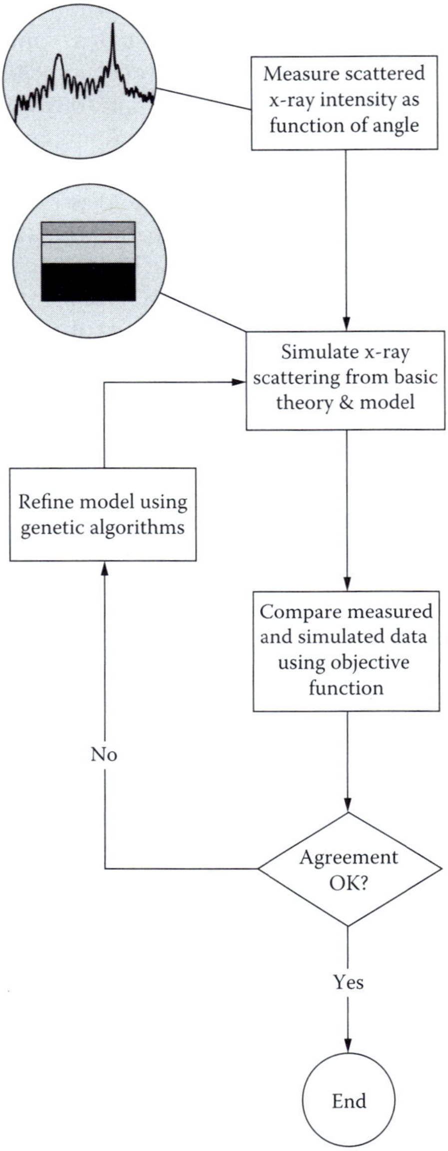

FIGURE 13.1
Flowchart for the differential evolution (DE) algorithm.

outside the specified constraints, it is replaced by a randomly selected value according to the expression

$$p'_j = p_j^{\min} + rand(p_j^{\max} - p_j^{\min}) \tag{13.2}$$

where p_j^{min} and p_j^{max} are the minimum and maximum permissible values of parameter j, respectively. The function rand(x) designates a real uniform random number drawn from the range (0, x). If the vector **t** satisfies the inequality

$$f(\mathbf{t}) \leq f(\mathbf{p}_0) \tag{13.3}$$

then **t** is selected to replace $\mathbf{p}_0$; otherwise, $\mathbf{p}_0$ propagates to the next generation. The procedure is then repeated for all remaining parameter vectors in the population **P**, that is, $\mathbf{p}_i$ with $i = 1, 2, \ldots, m - 1$, with a new vector **b**' calculated each time. Finally, the algorithm is iterated over many generations until $f(\mathbf{b})$ fails to decrease, i.e., until the best-fit vector **b** has converged on the global minimum of the objective function.

13.4.1 The Objective Function

The choice of an appropriate objective function is crucial for any data-fitting procedure regardless of the optimization method used. The DE algorithm gives us a great deal of flexibility in this choice since we need only choose a continuous function and do not require the function to have continuous derivatives (as is required by, for example, the Levenberg-Marquardt method). When fitting x-ray data, the objective function should have the following additional properties:

1. A single deep global minimum
2. Local minima that are much less deep than the global minimum
3. Fast and simple to calculate
4. Relative insensitivity to the absolute magnitude of the data
5. Does not overemphasize outlying points in the experimental data

Point 4 suggests that a logarithmic function could be appropriate since it linearizes data spanning several orders of magnitude. Point 5 suggests that a robust objective function[6] will be one that is more suitable than the mean squared objective function commonly encountered in least squares fitting. Tests of a number of objective functions that have been applied to fitting problems showed that the linear or least squared functions did not effectively fit the data at low intensities. These occur at large scattering angles and contain information on the smallest-length scales present in the structure, which are often those at which the x-ray characterization is aimed. However, the following could be satisfactorily used with high-resolution XRD (HRXRD) or XRR data:

Mean square difference of the log-transformed data:

$$f(p) = \frac{1}{N-1}\sum_{j=1}^{N}[\log I_j - \log I(\theta_j; p)]^2 \quad (13.4)$$

Mean absolute difference of the log-transformed data:

$$f(p) = \frac{1}{N-1}\sum_{j=1}^{N}|\log I_j - \log I(\theta_j; p)| \quad (13.5)$$

These could both cope adequately and generally with such data, but the second function is preferred because of its lower sensitivity to outlying data points (due mainly to statistical noise in the experimental data). Clearly we cannot assert that this is the best possible objective function, but it is very effective, and this is sufficient. In some cases faster convergence (and hence higher throughput) may be obtained by constructing an objective function closely adapted to the data obtained in a particular manufacturing process.

13.4.2 Performance and Examples

The performance of the data-fitting procedure is primarily affected by the following four factors:

1. The quality and size of the experimental data. If the experimental data are noisy or contain a very large number of points, it will take longer to determine the best-fit parameter values.
2. The search range of the parameter values. If a large search range is specified, the fitting procedure may take longer to converge to the global minimum of the objective function. However, because the DE algorithm is rather good at finding the global minimum, without becoming trapped in local minima, it tends not to falsely converge to incorrect values for the parameters.
3. The number of adjustable parameters. The ability of the procedure to determine the optimum parameter values decreases as the number of parameters increases. In practice, we find that up to 10 parameters can be optimized in a matter of minutes, and that several tens of parameters can be optimized during an overnight run.

As an example from XRR, we consider a Ta (10 nm) layer deposited on an Al_2O_3 substrate. Figure 13.2a shows the measured and simulated XRR curves before fitting. At very small angles of incidence, $\theta \le 0.5°$, the intensity is very high as a result of total external reflection from the Ta layer. As the incidence angle is increased, the reflected intensity decreases rapidly and prominent oscillations (Kiessig fringes) are clearly visible due to the interference of the waves partially reflected from the Ta layer and the underlying Al_2O_3

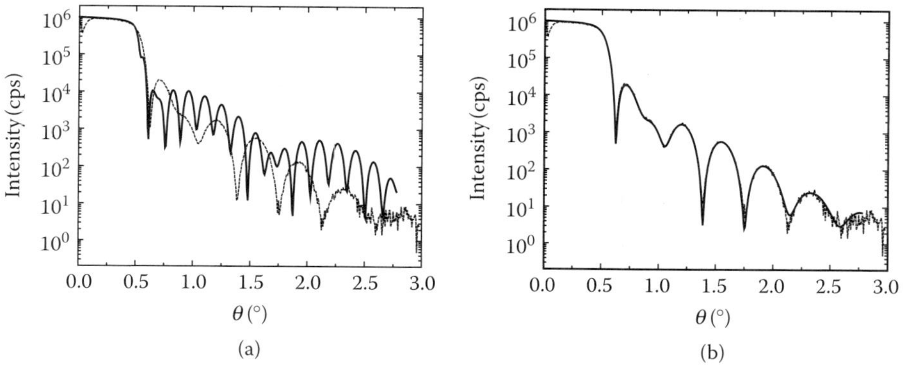

FIGURE 13.2
Comparison of experimental and simulated x-ray reflectivity curves for a Ta layer on Al_2O_3 (a) before and (b) after the fitting procedure has converged. The dashed lines represent the measurements, and the solid lines are the simulations.

TABLE 13.1
Best-Fit Parameter Values for the Ta Layer on Al_2O_3

Layer	Material	t (nm)	σ (nm)	ρ (nm)
2	Ta_2O_5	2.70 ± 0.05	0.71 ± 0.03	8.6 ± 0.2
1	Ta	10.49 ± 0.02	0.45 ± 0.02	16.1 ± 0.2
Substrate	Al_2O_3	∞	0.38 ± 0.02	3.99

substrate. The period, $\Delta\theta$, of the Kiessig fringes is related to the thickness, t, of the Ta layer according to the relation $\Delta\theta \approx \lambda/2t$. The amplitude of the Kiessig fringes depends on the difference in the refractive index of the layer and the substrate, which is quite large in this example. Figure 13.2b shows the measured curve together with its best-fit simulation. Nine adjustable parameters are fitted in a reasonable time. The best-fit parameter values and their uncertainties are given in Table 13.1. It should be noted that a surface oxide layer had to be included in the structural model to obtain close agreement of the measured and simulated curves.

The parameter values for the simulated curve shown in Figure 13.2a were chosen to be far from the anticipated best-fit parameter values. This was a deliberate choice in order to demonstrate that the fitting procedure rapidly converges to the global minimum in the objective function without getting trapped in local minima. The progress of the fitting procedure is illustrated in Figure 13.3, which shows the value of the objective function vs. the number of generations (iterations of the DE algorithm). Horizontal sections are times during which the fitting procedure is temporarily in a local minima. The fitting procedure is seen to have converged to the global minimum after only 1000 generations.

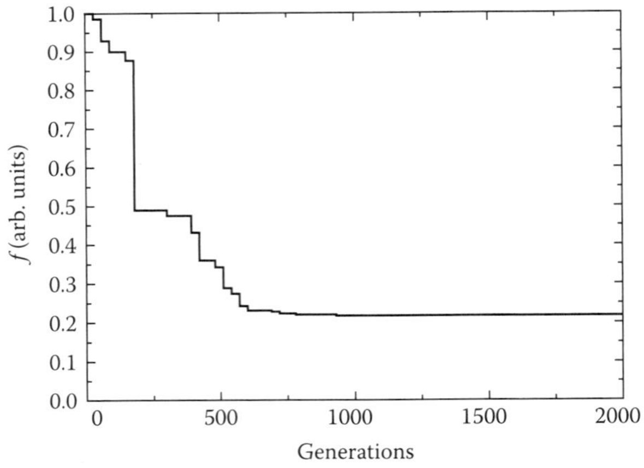

FIGURE 13.3
Variation of the objective function, *f*, with the number of DE generations, for the data-fitting process of Figure 13.2. The fitting procedure has converged after approximately 1000 generations in this case. (From M. Wormington et al., *Phil. Trans. R. Soc. Lond., A.* 357(1999), 2827–2848. With permission from The Royal Society.)

Figure 13.4 shows the value of the objective function as a function of the thickness of the Ta layer, with all other parameters held at their best-fit values. We note that the objective function has a single deep global minimum and many local minima. Harmonic minima, which occur at half and twice the best-fit Ta layer thickness, are the deepest of the local minima. The global minimum is shielded by fairly large maxima on either side. This is a typical feature in such curves when the thickness is varied, and turns out to be caused by the beating of two sets of oscillations (Kiessig fringes) in which

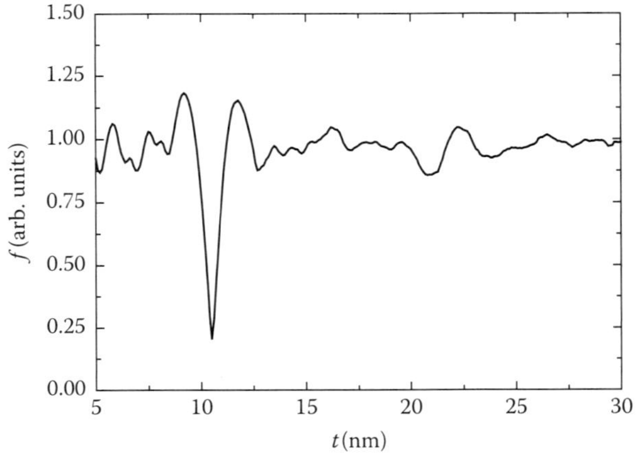

FIGURE 13.4
Variation of the objective function, *f*, with the Ta layer thickness. All other adjustable parameters in the model are held constant at their best-fit values.

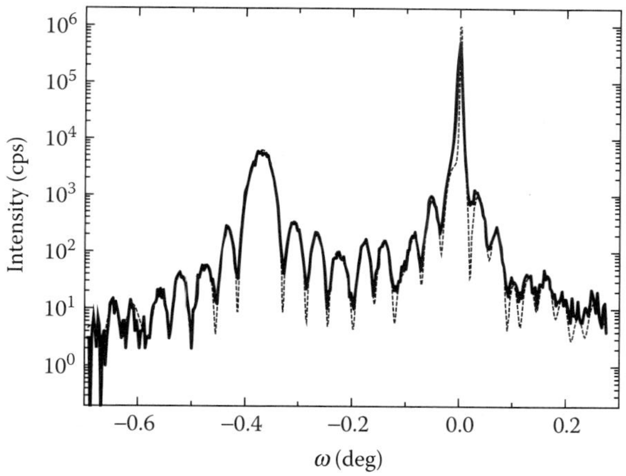

FIGURE 13.5
An XRD scan (solid line) of a box structure of SiGe (125 nm thick, 14.2% Ge) showing interference fringes, and the match (dashed line) obtained by automatic model fitting.

one period is fixed and the other is variable. This characteristic shape is very useful for recognizing whether the global minimum is in fact within the range specified for the thickness parameters in question. It also, incidentally, would slow down the convergence of the simulated annealing method.

An example from HRXRD of a 125-nm-thick 14.2% Ge layer of SiGe on Si is shown in Figure 13.5, which includes both experimental and simulated data. The objective function is shown in Figure 13.6. In the diffraction case, the large maxima either side of the global minimum are absent, but the essential features of many local minima plus a very strong, sharp, and deep global mimimum are identical.

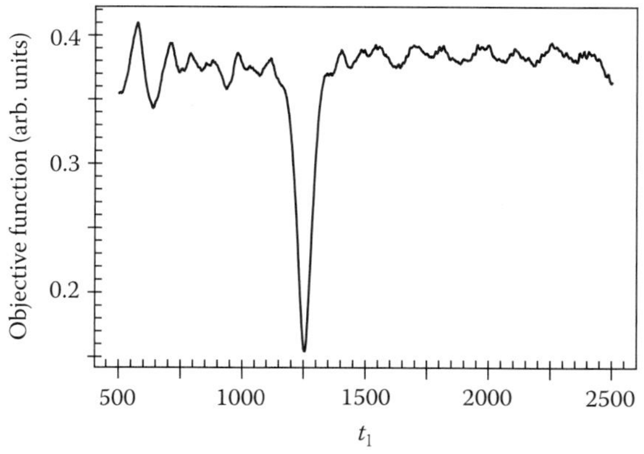

FIGURE 13.6
Objective function, *f*, for XRD scan of Figure 13.5. The thickness of the SiGe layer was varied, with all other parameters held constant at their best-fit values.

13.5 Requirements for Automated Analysis

Little needs to be done to adapt such algorithms to automated analysis; indeed, they are the core of this process. It is necessary to select an automatic stopping criterion. The choices are:

1. A specified number of simulations is reached.
2. A specified elapsed time is exceeded.
3. A specified value of the objective function is reached.
4. The user cancels the data-fitting procedure.

Criterion 3 is normally used in process development, but for automated fab analysis, criterion 1 or 2 is usually appropriate and leads to consistent cycle times. The criteria are set up during the recipe design, after appropriate process development, and depend upon the complexity of the film stack. It is normal to report the value of the objective function as well as the optimized parameters. If this is outside certain bounds, then it can be assumed that the material has failed the metrological test. It is more usual to find individual parameters, such as thickness or composition, out of control.

In all cases, it is necessary to transmit the resulting data to the factory control computer, following all appropriate SEMI (Semiconductor Equipment and Materials International) standards and protocols.

13.6 Summary

- X-ray theory can reliably predict the reflectivity and diffraction curves from wafers containing complex film stacks.
- There is no direct transform from data to structure, nor is there a uniqueness theorem. However, the known constraints in the manufacturing process practically eliminate ambiguity.
- Direct methods based on Fourier transforms are fast and may be used in simple cases to extract layer thicknesses, but are insufficiently general.
- Modeling methods provide a reliable and robust extraction of parameters in reasonable time.
- The differential evolution type of evolutionary algorithm has been found to be highly successful in parameter optimization.
- The modeling methods are easily automated by selection of stopping criteria and provision of outputs to appropriate SEMI standards.

References

1. J.M. Hudson, B.K. Tanner, and R. Blunt, *Adv. X-Ray Anal.*, 37 (1994) 135.
2. K. Sakurai and A. IIda, *J. Appl. Phys.*, 31 (1992) L113.
3. K.N. Stoev and K. Sakurai, *Spectrochim. Acta*, B54 (1999) 41.
4. R. Clinciu, M.Sc. thesis, University of Warwick, U.K., 1992.
5. I.R. Prudnikov, R.J. Matyi, and R.D. Deslattes, *J. Appl Phys.*, 90 (2001) 3338.
6. M. Wormington, C. Panaccione, K.M. Matney, and D.K. Bowen, *Phil. Trans. R. Soc. Lond. A*, 357 (1999) 2827–2848.
7. M. Wormington, K.M. Matney, and D.K. Bowen, Application of differential evolution to the analysis of x-ray reflectivity data, in *Differential Evolution: A Practical Approach to Global Optimization*, K. Price, R.M. Storn, J.A. Lampinen, Eds., Springer-Verlag, 2005.
8. M. Wormington, C. Panaccione, K.M. Matney, and D.K. Bowen, Fitting of X-Ray Scattering Data Using Evolutionary Algorithms, U.S. Patent 6,192,103, 2001.
9. P.R. Bevington, *Data Reduction and Error Analysis for the Physical Sciences*, McGraw-Hill, New York, 1969.
10. W.H. Press, B.P. Flannery, S.A. Teukolsky, and W.T. Vetterling, *Numerical Recipes in Pascal: The Art of Scientific Computing*, Cambridge University Press, Cambridge, 1989, chaps. 10 and 14.
11. M. Wormington, C. Panaccione, K.M. Matney, and D.K. Bowen, Automatic Parameter Optimization Software for X-Ray Reflectometry and Diffraction, U.S. Patent 6,192,103, granted 2000.
12. K.V. Price and R. Storn, *Dr. Dobbís.*, April 1997, pp. 18–24.

14

Instrumentation

14.1 Introduction

In this chapter we describe the technology currently used in x-ray metrology (XRM) tools. Since this changes rapidly, it is given in outline only, simply in order to give some familiarity with the tool systems used in practice. We emphasize the fundamental issues of metrology in x-ray tools. As has been seen (and is discussed further in Chapter 15), the interferometric methods of XRM are traceable with suitable care. They depend upon knowledge of the wavelength and angle metrology, and we shall point out how these are determined. The implications on repeatability, reproducibility, and accuracy or trueness are discussed in Chapter 15.

14.2 X-ray Sources

An x-ray source is required to deliver a beam of well-defined wavelength, sufficient intensity, and small enough size. The last is now required to be below 100 μm at the sample so that measurements can be made on wafers in the scribe lines between dies.

The only current practical sources for x-rays in fab tools are electron beam sources with water-cooled targets. The most useful choices for target are copper and molybdenum, with wavelengths of 0.1540 592 90(50) and 0.0709 317 15(41) nm, respectively. These are experimental values, traceable by NIST to the international meter to the specified precision. This is better than 1 in 10^8, some 10,000 times better than is required for reproducibility or accuracy in XRM.

While rotating anode sources are powerful, they are avoided if possible in fabs because of their high maintenance requirements and the need for 95% uptime throughout the year. Microbeam sources, which give the highest brightness and are best adapted to advanced x-ray optics, are the

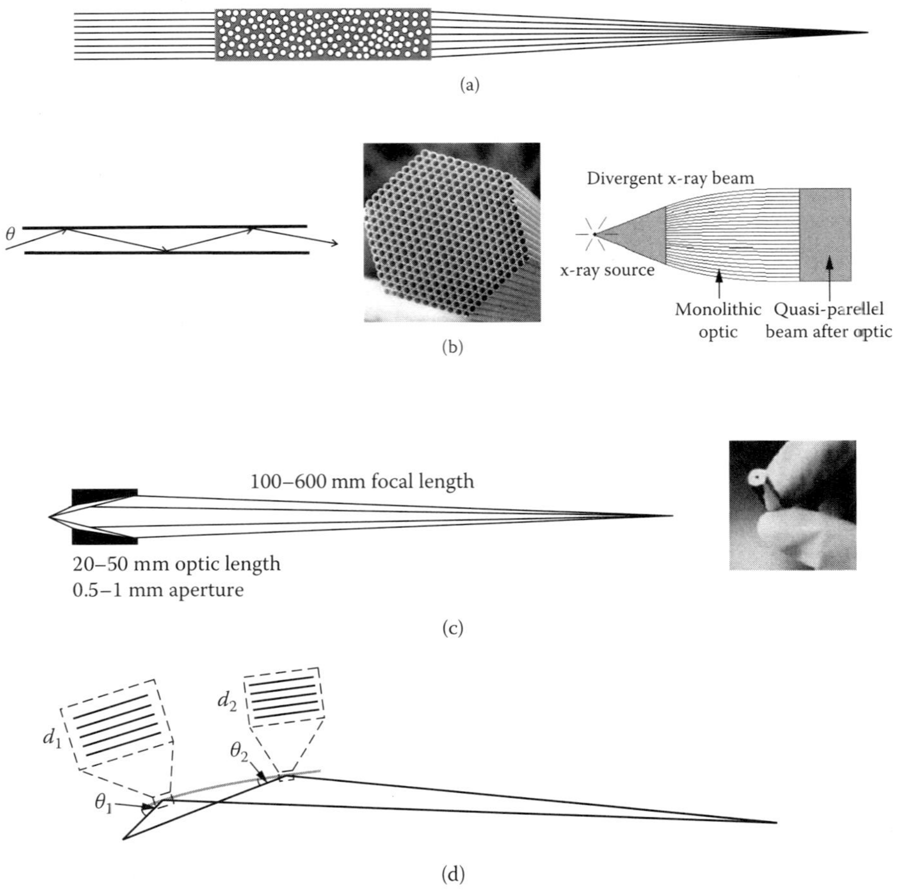

FIGURE 14.1
The principles of x-ray optics. (a) Refractive optics. (b) Specular optics. (Courtesy Reflex s.r.o.) (c) Polycapillary optics. (Courtesy XOS, Inc.) (d) Multilayer optics.

usual choice. Brightness is the number of photons per second in a given beam size and in a given angular divergence, measured in photons sec^{-1} mm^{-1} $steradian^{-1}$. Brightness increases as the electron beam source size decreases, since the cooling of the spot becomes more efficient. Moreover, small spots are required to utilize focusing optics (see below). The Liouville conservation theorem states that brightness is constant throughout a (lossless) optical system. Hence, if a small spot with low divergence is required at the specimen, there is no choice but to start with a small, bright spot at the source.[1]

14.3 X-ray Optics

In recent years, the old adage that x-rays cannot be focused has been proved false. There are four main means of focusing an x-ray beam, illustrated in Figure 14.1:

- **Refractive optics**:[2] Since the refractive index of materials for x-rays is less than unity, these consist of low-absorbing materials containing voids. The void acts as a true refractive lens. However, these are best used at much higher energies (shorter wavelengths) than are common for XRM, because of absorption losses. At high energies the following solutions are less effective.
- **Specular optics**:[3] Since x-rays undergo total external reflection at small angles, a grazing incidence ellipsoidal mirror will focus an x-ray beam from one of its foci to the other. An ellipsoidal mirror (shown) will focus a beam from one focus of the ellipse to another, and a parabolic mirror will collimate a beam starting from its focus. These optics are best at lower energies/longer wavelengths, for example, for x-ray lithography or astronomical x-ray optics.
- **Polycapillary optics**:[3] A gently curved hollow capillary will act as an x-ray waveguide if the curvature is low enough that the internal reflections are kept below the critical angle. A shaped bundle of capillaries can either focus or collimate a beam. These optics work well at wavelengths used in XRM, though the minimum divergence of the beam is approximately 0.2°. A very good application is for a small, focused beam for x-ray fluorescence (XRF) analysis, for which the divergence is unimportant.
- **Multilayer optics**:[3] The geometry of these is similar to specular optics. However, to increase the acceptance angle and aperture of the optics, the surfaces are coated with multilayers of dense/light bilayers, such as Mo-Si. These give Bragg reflections at up to a few degrees, and excellent control of divergence. However, the technological demands are extreme, since the surfaces are aspheric in both axes, need submicron precision and subnanometer roughness, and the multilayer spacings have to be graded in period from one end of the optic to the other, so that the Bragg law is obeyed at all points of the surface as the angle of incidence varies. The ellipsoidal, focusing geometry is illustrated.

A state-of-the-art example, which gives independent focal length or collimation control in two axes, is shown in Figure 14.2. This is used to give high intensity in sub-100-μm spots (for scribe line metrology) while preserving

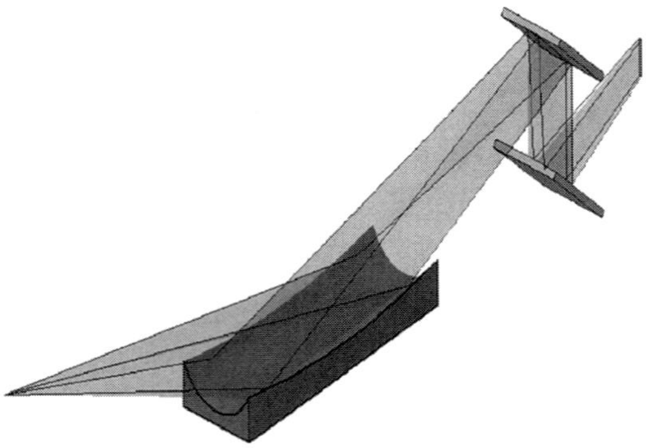

FIGURE 14.2
A dual-focusing optic (the FOX optic from Xenocs). The beam is collimated in the incidence plane to maintain high resolution, and focused in the normal plane to increase intensity in a small measuring spot. (Courtesy Peter Hoghoi, Xenocs s.a.)

high angular resolution for x-ray reflectivity (XRR) or high-resolution x-ray diffraction (HRXRD).

Multilayer optics also provide monochromatization, in the range 10^{-2} to 10^{-3} $\delta\lambda/\lambda$, which is usually sufficient for XRR and XRD and certainly for XRF. HRXRD, however, requires a narrower spectral bandwidth, down to about 10^{-4}, and crystal optics are required in addition. These work on the same principle of Bragg diffraction that was discussed in depth in Chapter 11. There it was noted that plane wave rocking curves of crystals are of the order of a few arc seconds wide. Using the differential form of the Bragg law,

$$\delta\lambda/\lambda = \cot\theta\;\delta\theta \tag{14.1}$$

We see that 5 arc sec width at a Bragg angle around 44° will give a bandwidth of 10^{-4}. In practice, we can tune the bandwidth and the collimation over a range of about 100× by choice of reflection, material, and symmetry of diffraction. The principles are:

- Increasing the atomic number Z of the material broadens the rocking curve and thus increases bandwidth. In practice, Si and Ge are the normal choices through utility, quality, and availability.
- The larger the structure factor, the larger the bandwidth and the broader the curve.
- The larger the asymmetry of the reflection in grazing incidence, the larger the bandwidth accepted and the narrower the bandwidth transmitted by the crystal.
- The large tails of the rocking curve can be sharply reduced by using multiple parallel reflections from the perfect crystal (Figure 14.3).

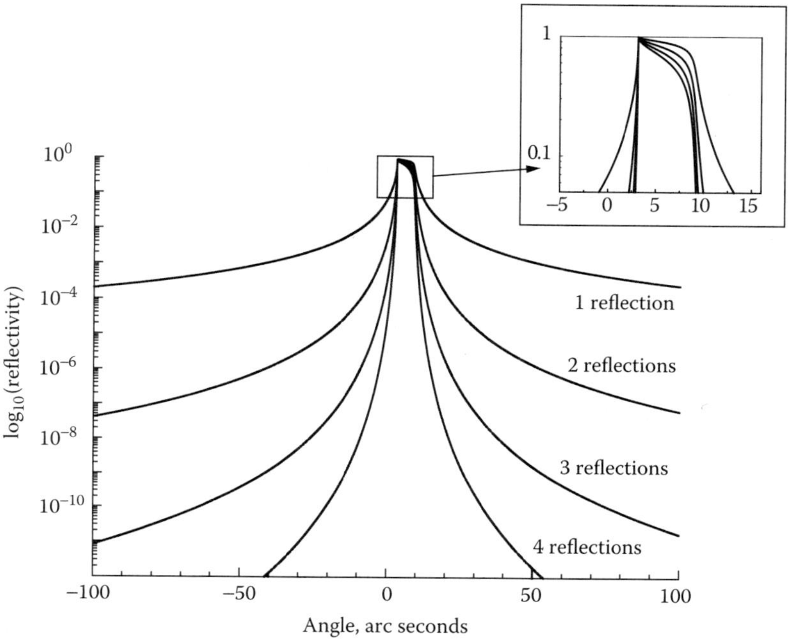

FIGURE 14.3
The rocking curves produced from one, two, three, and four successive reflections in a channel-cut Si 220 crystal for CuK radiation. The inset shows that little intensity is lost near the peak, but the tails are greatly reduced.

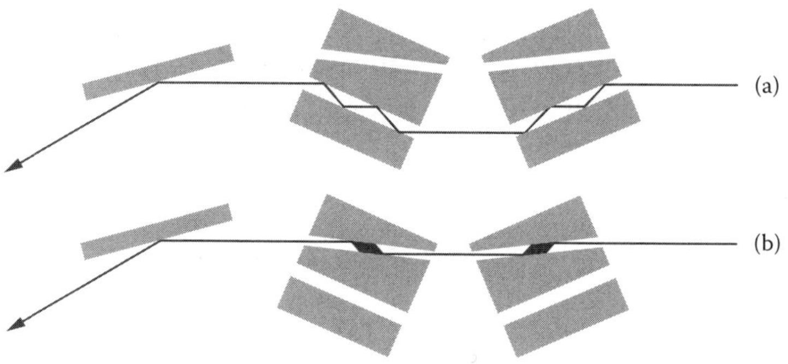

FIGURE 14.4
The Loxley–Tanner–Bowen[5] combined high-resolution and high-intensity beam conditioner based on the duMond principle, illustrated for Si 220 with CuK1 radiation. (a) Geometric arrangement for high resolution. (b) Geometric arrangement for high intensity.

Many designs of crystal monochromators are available; for a review and discussion of the theory, see Bowen and Tanner.[4] An example of a beam conditioner that controls both wavelength and collimation at low- and high-resolution levels is shown in Figure 14.4.

14.4 Mechanical Technology

14.4.1 Angle Measurement and Calibration

There are two different approaches to angle measurement. The most fundamental and the only traceable method is to use a goniometer with encoded axes. Encoders are normally optical, and it is essential to use a complete 360° encoder. This may then be calibrated by the manufacturer, either in manufacturing a certain number of divisions per full circle or by counting them after manufacture. The overall calibration is then referred to a natural standard of the full circle. The accuracy of division into fringes (linearity) is then the only issue. It is eased because readout heads are interferometric and average a number of divisions at each reading. For the most accurate and precise metrology, a number of read heads around the circle are used. If there are N heads, the harmonic errors per revolution will be N-fold. These arise from eccentricity of mounting the encoder and, if the axis is mounted on a ball or roller bearing, from minor variations in the ball or roller bearing diameter and their precession about the circle. However, with current machining, encoder, and bearing qualities, it is straightforward to obtain linearities and hence traceable angular measurements to a few arc seconds in a revolution, say 1 in 10^5, with no special precautions. This is at least 10 times better than the required reproducibility for fab material measurements.

Linear or area detectors (see Section 14.5) are also used. Here the signal is dispersed in space, for example, using a focusing polycapillary optic, and the plot of scattered intensity vs. angle is reconstructed from a linear or area detector spanning the range of the measured angles. These are not directly traceable, nor, usually, are they adequately linear. They therefore require calibration at each point on their scale. They offer somewhat faster data collection than the scanning methods at the expense of poorer reproducibility between tools. For some x-ray methods, such as polycrystalline texture analysis, grain size measurement, and mosaic metrology in general, this can be a very good trade-off. The absolute accuracy needed for such methods is not nearly so high as for XRR or XRD.

14.5 Detectors

Since there is rarely as much x-ray signal as would be desirable, it is essential that x-ray detectors be photon counting. A very high dynamic range is also required, since the metrology signal may be many orders of magnitude weaker than the substrate signal. This is achieved in a number of ways.

Single-point detectors comprise scintillation counters, proportional counters, ion chambers, and silicon-based sensors (PIN diodes or SDDs). Ion chambers are generally not photon counting, and proportional counters saturate at about 10^5 cps. Silicon sensors have the lowest noise and are used for efficient measurement of weak signals, and also for XRF scatter since they have good energy resolution. Scintillation counters can give dynamic range from 0.2 to 10^8 cps and are in frequent use. Some silicon sensors can achieve a very high dynamic range, but are more often restricted to $<10^5$ cps.

Linear and area detectors may be based on silicon sensor arrays or on wire proportional counters. Sometimes fiber-optic coupled phosphors must be used to gain enough signal, and in these cases spatial linearity must be calibrated. High-speed data acquisition may be achieved with such detectors, since they eliminate one or two mechanical scans. However, strict calibration routines are necessary for reproducibility.

A further advantage of two-dimensional detectors is the ability to post-process the data. An outstanding example of this is the digital defect mapping method discussed in Chapter 12, in which small diffraction stripes are integrated to form a full wafer map. Even in HRXRD or XRR this can be useful. The signal received at the detector is a mix of the desired signal and the noise. In conventional measurements, one uses physical slits to eliminate the noise. With an area detector one may simply select the region of interest (ROI) corresponding to the slit width. Postprocessing may be used to alter the "virtual slit" width to optimize the signal/noise for a given application.

14.6 Practical Realizations

In Table 14.1 we summarize the specifications required for tools to perform x-ray metrology by the major techniques of XRD, HRXRD, RSM, XRR, defect mapping, and diffuse scatter.

Finally, we give illustrations of commercial instruments that perform XRM in a fab environment, whether for research or in production (Figure 14.5). An illustration of a diffraction mapping (defect imaging) tool was shown in Chapter 12.

14.7 Summary

- Traceable XRM depends only on the accuracy of measurement of wavelength and angle. The former is traceable to true measurements to 1 in 10^8, and the latter to 1 in 10^5 with no special precautions and to 1 in 10^7 if required.
- Microbeam sources and focusing x-ray optics are in common use for fab tools. They have proven stability and reliability in the fab.

- High-dynamic-range point detectors give the highest reproducibility and accuracy, but linear and area detectors are sometimes used to increase throughput.
- Postprocessing of area detector data may be performed to improve signal/noise.

TABLE 14.1

Specifications for Instruments for XRM by the Techniques Discussed in this Book

Technique	ω Range	ω Precision	2θ Range	2θ Precision	Dynamic Range of Detector
XRR	0–4°	0.001°	0–8°	0.001°	10^8
XRR diffuse scatter	0–4°	0.01°	0–10°	0.01°	10^4
XRD	0–90°	0.01°	0–110°	0.01°	10^6
HRXRD	0–90°	0.0005°	0–110°	0.0005°	10^4
RSM	0–90°	0.0005°	0–110°	0.0005°	10^6
GIXRD	0–4°	0.1°	0–110°	0.01°	10^4
GIIXD	0–90°	0.05°	0–110°	0.05°	10^4
Defect mapping	0–50° reflection 0–30° transmission	0.01°	0–90°	0.1°	10^3

Note: Ω is the angle of the incident beam on the specimen, and 2θ is the angle of the scattered radiation with respect to the incident beam.

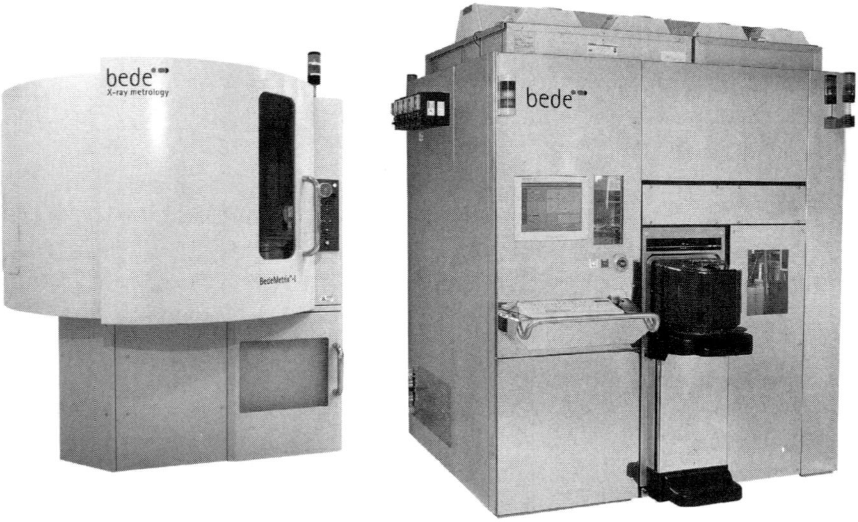

(a) (b)

FIGURE 14.5
A combined XRR-XRD-HRXRD tool for (a) fab research applications and (b) in-line production control. (Courtesy Bede Scientific.)

References

1. U.W. Arndt, J.V.P. Long, and P. Duncumb, *J. Appl. Cryst.*, 31 (1998) 936–944.
2. B. Lengeler, C.G. Schroer, M. Kuhlmann, B. Benner, T.F. Günzler, O. Kurapova, F. Zontone, A. Snigirev, and I. Snigireva, *J. Phys. D Appl. Phys.*, 38 (2005) A218–A222.
3. A. Michette, *Optical Systems for Soft X-Rays*, Plenum Press, New York, 1986.
4. D.K. Bowen and B.K. Tanner, *High-Resolution X-Ray Diffractometry and Topography*, Taylor & Francis, London, 1998.
5. N. Loxley, B.K. Tanner, and D.K. Bowen, *J. Appl. Cryst.*, 28 (1995) 314–317.

15

Accuracy and Precision of X-ray Metrology

15.1 Introduction

In Chapter 1 we defined x-ray metrology (XRM) as the application of x-ray interference for measurement of material dimensions. This is achieved by splitting the x-ray wavefront through scattering by the material, and detecting the recombined component waves. The two fundamental parameters measured through such interference are layer thickness (in any structure, including amorphous) and strain (in crystals). Parameters that may be reliably derived from these include composition in crystalline materials, mosaic spread, and lateral nanostructure. We shall show in this chapter that such parameters can be measured or deduced with full traceability to natural or international standards.

A further class of material parameters may be inferred from x-ray data by use of some model or distribution function or other assumption. These include porosity, pore size distribution, grain size, roughness, rugosity, and x-ray fluorescence (XRF) measurement of thickness. While these can be of great importance, they are not fundamentally traceable, and are treated separately in this chapter.

Metrology for semiconductors has reached a critical stage. New materials, such as SiGe epitaxial transistors and barrier-isolated damascene copper conductors, are already in production, and others, such as strained silicon and high-*k* dielectrics, are at an advanced stage of development. Some of these materials are required to be nonuniform, for example, graded compositions of SiGe or porous dielectrics, or deposited as extremely thin layers. Metrology of these new materials is very challenging, but essential for process development and monitoring. The silicon industry is now coming to terms with the need for absolute accuracy or trueness of metrology, defined as a traceable unbroken chain of measurement from the device layer to natural or international standards, with a precision that does not degrade the repeatability. This is a new concept for fabrication lines that have historically relied on "tool matching" to get multiple copies of a vendor's systems to perform in the same way. Some factories even define "vendor" angstroms that apply to a specific model of tool. However, the absolute error in a vendor-defined unit may now exceed the size of the parameter being controlled, as layers approach atomic thicknesses. It is for this reason that advanced manufacturers are attempting to use *accurate and true*, not merely precise and repeatable, metrology systems in new factories.

In the near future it will become impossible to ensure long-term drift-free stability for semiconductor metrology by relying on "golden copies" of tools or on any standard materials other than intrinsic standards, such as the lattice parameter of silicon of specified purity, or international standards. However, x-ray analysis can be made traceable, since it depends upon two simple parameters that are themselves traceable: wavelength and angle.

15.2 Design of X-ray Metrology

In any metrology there are limitations and trade-offs. The main limitation for x-ray metrology is the limited brightness (photons sec^{-1} mm^{-2} $steradian^{-1}$) of portable x-ray sources, compared with optical sources. The Liouville theorem shows that brightness, the product of the beam size and its angular divergence, remains constant through any lossless optical system, whatever the optics used in the beam path. In practice, it will decrease due to absorption and scattering losses. The statistical error in a photon source is governed by Poisson statistics; i.e., the variance is equal to the total count in the measurement. A consequence of these laws is that the data collection time, the repeatability, and the spot size are coupled. Only two of these may be optimized at will, and the third will be dependent on the source brightness; however, repeatability can always be improved by increasing the data collection time.

Recent advances to improve the base of the trade-off include high-brightness microfocus sources and precision-graded multilayer x-ray optics that utilize these effectively with low losses. We are currently able to measure a film of thickness ~30 nm in x-ray reflectivity, within a 100-μm scribe line (street), to 0.65% 1σ at 150 wafer sites/h, and to 0.33% 1σ at 75 sites/h.

15.3 Repeatability and Reproducibility

All the statistical terms used are carefully defined in ISO 5725-1. The repeatability tells us how similar are a series of measurements of the same quantity, when the following conditions are constant:

- The operator
- The equipment
- The calibration of the equipment
- The environment
- The time between measurements, which should be *as short as possible*

Repeatability is expressed as the standard deviation of the sample.

The reproducibility of a measurement is the same quantity measured when the repeatability conditions are not maintained.

These two terms form the lower and upper limits of the concept known as precision. Repeatability tells us the best precision that the tool can achieve. Reproducibility tells us the worst precision. Tool matching in a semiconductor fab is appropriately referred to as a specification on the reproducibility of a tool.

In the x-ray case, the precision is largely determined by the following:

1. The absolute x-ray signal received by the detector. In general, the repeatability will improve as the square root of this signal.
2. The sensitivity of the x-ray signal (e.g., fringe contrast) to the parameter being measured. The repeatability will improve roughly linearly with this sensitivity.

It should be emphasized that rigorous setup, alignment, and checking procedures must be followed to eliminate errors relating to the alignment of the tool and the sample. For example, when testing a tool on a standard sample, it is often necessary to measure the sample at the same physical location within 20 μm to avoid errors due to variability in the sample.

The repeatability and reproducibility do not take account of any systematic offset between the measured and true values of the parameter being measured.

15.4 Accuracy and Trueness

Accuracy and trueness are measures of how close is the measurement, or measurements, to the real value. ISO 5725-1 defines *accuracy* as the closeness of a single measurement and *trueness* as the closeness of the average of a large set of measurements. The discrepancy between the reading (or average of a set of readings) and the true value is called the bias of the measurement(s).

In the following sections we discuss the implications for both x-ray reflectivity and x-ray diffraction measurements. It is assumed that we employ the best model-fitting routines as described in Chapter 13.

15.4.1 X-ray Reflectivity

X-ray reflectivity (XRR) is the measurement of the specular reflectivity at grazing incidence, typically 0 to 4° (see Chapters 1, 2, and 8). Since a grazing incidence method is used, it is not practicable to use a small spot for analysis. However, a narrow line, say 100 μm × 2 to 5 mm can be used, and XRR measurements may be performed in a scribe line between dies on a patterned wafer. The thickness of a single layer is directly calculable from the spacing of fringes in the reflectance spectra. For multiple layers, the method of model fitting using genetic algorithms (Chapter 13) has turned out to be extremely powerful and largely deskills this critical operation.

The use of x-ray interference fringes for determination of thickness illustrates an important feature of x-ray characterization methods: *they do not depend upon the parameters of the material being characterized.* Optical methods such as ellipsometry require that the state of the material itself (its refractive index) be part of the model. But the x-ray refractive index of *all* materials differs from unity by less than 0.01%. The traceability of the modeling process thus depends only on the traceability of angle and input wavelength, and of the relative reflectivity.

Angle is self-referencing to a complete circle (2π). Linearity of interpolation is normally handled to sufficient accuracy (50 μrad) by the manufacturers of the optical circle encoders used on the reflectometer axes. Calibration methods to 0.1 μrad[1] have been developed by NIST in a project to produce secondary standards to check any x-ray tool. *Input wavelength* of the characteristic x-radiation used is linked traceably to the international meter by NIST to an accuracy and precision better than 0.0001%.

15.4.2 High-Resolution X-ray Diffraction

In high-resolution x-ray diffraction (HRXRD), the diffraction profile is determined with high angular resolution (Chapters 1 to 4, 10, and 11). Since incident angles are typically tens of degrees, small spots (say, 100 × 100 μm) can be used, giving access to features such as selected-area epitaxy on patterned wafers.

The thicknesses of the layers are again determined from the interference pattern, and this is traceable to the international meter on the same arguments given for XRR. The only additional parameter is the Bragg angle (diffraction angle) of the diffraction plane used, which for silicon is known traceably to better than 0.00001%. The strains both normal and parallel to the interface can be measured, leading to a direct measure of lattice relaxation in relaxed epilayers and of the strain in strained silicon.

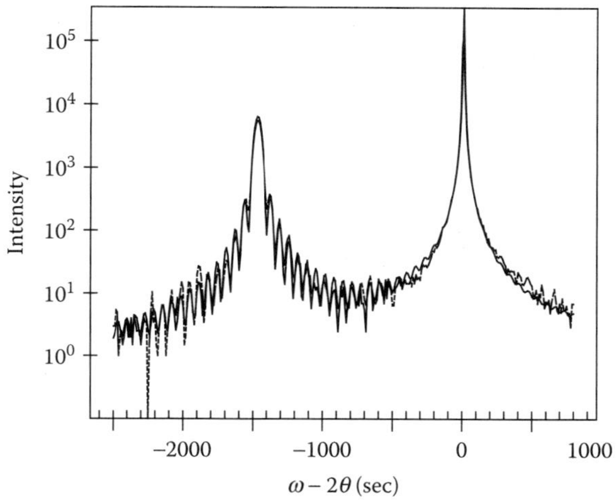

FIGURE 15.1
HRXRD curve of SiGe (dashed) and its modeled curve (solid) for the sample used for the gauge study. (The specimen was kindly provided by Antonio Terrasi, University of Catania, Italy.)

Composition is not measured directly but is deduced from the strain that (in this case) the Ge introduces to the Si lattice. For a compound of unknown composition this is assumed to be a linear interpolation between the lattice parameters of the constituent elements (Vegard's law), which, if not eliminated by calibration, may introduce a degradation in traceability of up to about 2%. However, accurate calibration figures are available in the literature for the case of SiGe at all required concentrations, and it is in any case the strain itself that is the more important parameter in device performance.

15.5 Repeatability and Throughput

In order to quantify the effect of Poisson statistics and data collection time upon the repeatability, we report a study[2] on a SiGe single layer, nominally 300 nm thick, and 16% Ge by HRXRD (Figure 15.1).

Ten repeats of data sets were taken, with count times from 0.1 to 30 sec per angle step. All data sets were then compared by a fitness function similar to that used to find the difference between experimental and simulated data in automated fitting procedures (Chapter 13). The two curves n and m have an error or difference factor given by

$$E = \frac{1}{N-1}\sum_{j=1}^{N} |\log I_j^n - \log I_j^m|$$

TABLE 15.1

Error E between Data Sets at Each Counting Time, Compared with Repeatability of the Deduced Parameters Thickness, *t*, and Composition, *X*

Count Time (sec)	E	Average *t* (nm)	% σ	Average *X*	% σ
0.1	0.69	298.9	0.63	0.15836	0.11
0.5	0.26	297.1	0.23	0.15862	0.24
1.0	0.16	297.2	0.28	0.15826	0.08
2.0	0.11	296.9	0.13	0.15886	0.20
5.0	0.06	297.1	0.14	0.15881	0.04
10.0	0.04	297.0	0.10	0.15887	0.03
30.0	0.06	297.0	0.13	0.15844	0.33

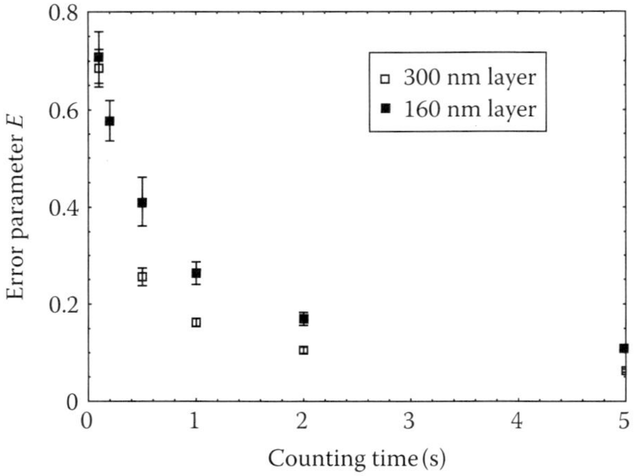

FIGURE 15.2
The effect of counting statistics on the error in determination of thickness.

in which *N* is the number of points in the curve. The results are shown in Table 15.1.

As expected, the counting statistics have little effect on the composition, which is determined by the precision of location of the large layer peak, but they do have a significant effect on the thickness. This is shown graphically in Figure 15.2. It is interesting that there appears to be a linear relationship between the repeatability and the error parameter — recalling that the latter simply expresses the similarity between data sets taken on the same specimen (see Figure 15.3). This gives a means of predicting the effect upon repeatability of variation in the data acquisition time.

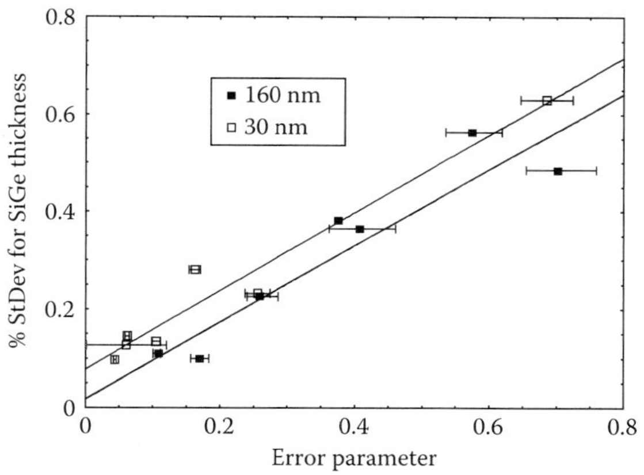

FIGURE 15.3
The relationship between the repeatability of the SiGe thickness and the error parameter.

15.6 Absolute Tool Matching

If metrology tools are referenced to absolute standards, and correct procedures uniformly applied, the results should be not only repeatable but reproducible, and all tools should automatically match each other. This assertion has been tested extensively on tools shipped to XRM customers. An example for metrology of a SiGe layer by HRXRD is shown in Table 15.2.

If the angles are properly referred to an absolute measurement, in no case is it necessary to use one tool as a golden standard, and no matching factors are required in the software on any tool. The only parameters unique to each tool are the encoder calibration (encoder divisions per 360°, supplied by the manufacturer) and the offset values of the datum signals of the encoders (determined during the alignment procedure). It is essential to develop rigorous alignment procedures, but tool matching is then indeed automatic.

TABLE 15.2
Comparison of Two Tools

	Tool 1	Tool 2
Thickness (nm)	47.34	47.44
1σ (nm/%)	3.5/0.74	3.8/0.8
Composition (%)	19.95	19.94
1σ (abs/%)	0.04/0.2	0.05/0.25

Note: The figures show matching to 0.2% in thickness and 0.05% in composition, with no adjustment or tool-matching parameters.

The concepts here demonstrated with SiGe film thickness by HRXRD also apply to thin films such as barriers associated with Cu interconnects, measured with XRR.

15.7 Specimen-Induced Limitations

15.7.1 Effect of Layer Defects

X-ray analysis may not apply equally well to all materials. Materials with high contrast between layers (e.g., HfO_2 on Si) can be measured easily down to 1.0 nm, but measurement is more difficult and repeatability poorer for materials (e.g., SiO_2 on Si) that show lower contrast. Sample-dependent processes can also affect the measurements, e.g., excessive interface roughness or dislocation content. The repeatability, accuracy, and fitness function for poor wafers will be much poorer than for good wafers, but this in itself is an indication of quality. Our experience is that in the early stages of process development the material control and the repeatability may be relatively poor. If the fitness function is significantly poorer than expected for the film stack/material type, this is in itself an indication of material quality. As the process control improves, so does the repeatability and fitness function. The reason is that, as in any interferometric measurement, the quality of the surfaces and uniformity of the films have a direct effect upon the signal/noise in the measurement. The signal/noise is thus a function not only of the tool, but also of the film stack on the sample.

15.7.2 Where Is the Surface?

The thickness of layers that must be measured and controlled in semiconductor technology is now approaching atomic dimensions. The isolating Cu layer in spin valve magnetic structures and the high-k gate oxide in transistor structures may be only 1 nm thick. Measurement repeatabilities demanded by semiconductor manufacturers are typically <1% even at these thicknesses. In view of the discrete nature of atoms, with diameters around 0.3 nm, the question is often asked: What does a thickness tolerance of 0.01 nm actually mean? And where is the surface?

The relevance of the latter question is seen by the thought experiment indicated in Figure 15.4. Let us assume that the composite surface is by some means made perfectly flat, which we may define, say, as coplanarity of the atomic nuclei in the surface layer. These are sufficiently small that they can be considered as points on our scale of measurement. Now consider how different probes will measure the surface, noting that they mostly will interact with the electron clouds rather than the nuclei:

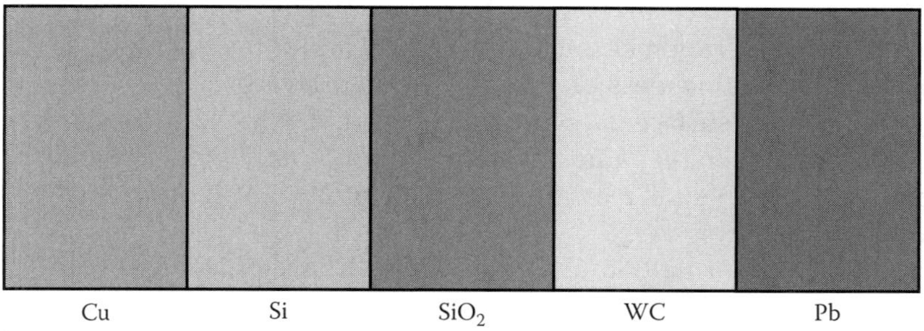

FIGURE 15.4
A composite material for a thought experiment — where is the surface and where will metrology tools say that it is?

- An atomic force microscope (AFM) will measure the compliance of the surface atoms. It will sink in farther to a large atom such as lead than it will with a small, tightly bound atom such as silicon. The transition between layers will be sharp and influenced by the tip radius.
- A scanning tunneling microscope (STM) will measure the local work function of the material. It will be different for each conductor and simply produce noise for insulators. The transition between layers will be sharp, related to the tip radius but differently from AFM.
- An optical probe used in reflection will interact with approximately the top 5 to 10 nm of surface, depending on its transparency. Polarization and phase are slowly changed by the transit through this region. The transition between layers will be gradual, related to the wavelength and the beam size.
- An x-ray probe used in total external reflection will sample 2 to 3 nm of material, and in diffraction up to 10 μm. The penetration will be lower for materials with higher atomic number, which scatter the beam more strongly. Polarization is little affected by the material, but phase is strongly affected. The transition between layers will be gradual, related to the wavelength and the beam size.
- A neutron probe will interact directly with the nuclei, but it will interact more strongly and penetrate less deeply into materials with higher cross sections for nuclear scattering. The transition between layers will be gradual, related to the wavelength and the beam size.

A scan of each of these across the composite material will produce significantly different results. Each will show a discontinuity, of varying height and abruptness, at the junction between materials. It is not, therefore, surprising that different methods give different data for layer thicknesses.

An epitaxial surface is quantized in atomic layers and may be partially covered. Some metrologies, including optical and x-ray, will give some sort of average of the thickness of partially covered layers.

An amorphous surface is not quantized and, if it has sufficient thickness, can have any thickness value above one atom thick. Models of amorphous layers show that small rearrangements of atoms can accommodate very small changes in thickness.

The surface is normally much flatter than the size of its constituent atoms might suggest. The depressions between atoms are only a small fraction of an atomic diameter when smoothed by the electron cloud around adjacent atoms. Only the STM and AFM can distinguish such ripples.

15.7.3 Comparisons of XRM with Other Metrologies

There is some information available on comparison of XRM with other methods. Figure 15.5 gives the results of a round-robin organized by NIST, showing that XRR agrees well with the mean of other methods, including spectroscopic ellipsometry (SE). The work of Kohli et al.[3] provides a direct comparison between XRR and SE and also summarizes earlier work. In general, the data from SE and XRR are linearly related with a slope very close to unity, but SE data are generally offset in a positive sense (layers appear thicker in SE) with an intercept between 0.27 and 1.7 nm. This probably depends upon the interface layer between the substrate and film, and on the surface contamination to which SE is more susceptible. Both effects would increase the apparent measured thickness. XRR would therefore seem to be significantly more reliable for films below 2-nm thickness.

In XRD some studies have been performed on measurements of Ge content by XRD and SIMS. A single structure was examined in detail, in a joint

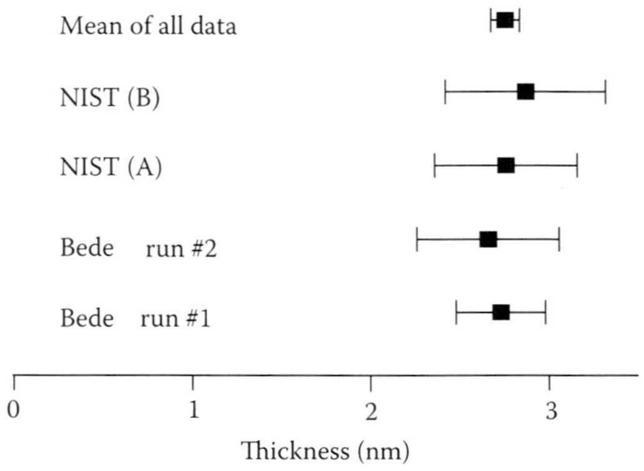

FIGURE 15.5
NIST round-robin on oxynitrides: a comparison of data from two laboratories on sibling samples. The "mean of all data" includes all measurements, such as spectroscopic ellipsometry.

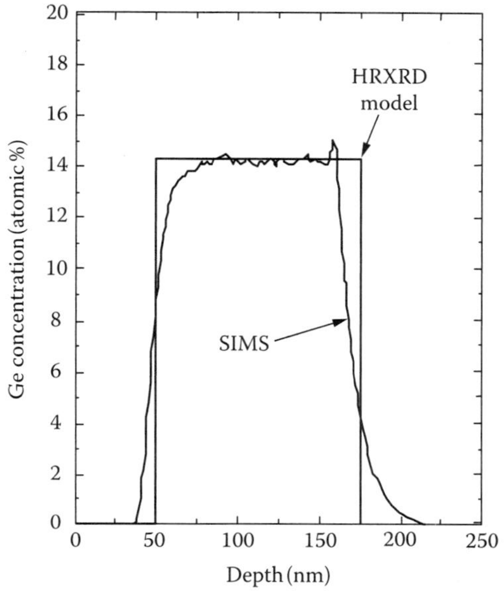

FIGURE 15.6
SIMS measurement of the sample used for the HRXRD data in Figure 1.11 (courtesy Dr. Annelena Thilderquist, Applied Materials) and the box profile deduced from XRD. The slope errors on either side of the box are ascribed to the well-known beam-mixing artifacts in SIMS. Note the excellent agreement on the steady-state composition away from the edges of the box.

assessment between Applied Materials, Inc., and Bede (see Figure 15.6). Agreement is excellent. A series of box structure Si-Ge samples was also analyzed for Ge content by HRXRD at Bede and by a commercial SIMS laboratory (see Figure 3.6). Correlation between all the independent SIMS and HRXRD results was 99%.

The arguments in Section 15.7.2 indicate that we should never expect perfect agreement between physically different measurements. However, the reliability in practice of x-ray measurements down to at least 1-nm thickness does appear to confirm the arguments presented earlier on absolute, traceable metrology.

15.8 Summary

- With technology ramp rates accelerating and profitability windows narrowing, it is not enough to monitor run-to-run variability in the fab. Access to accurate information allows faster development and troubleshooting.

- The methods of XRM not only have the sensitivity required for the analysis of novel materials (at least to the 32-nm node), but are also accurate.
- True measurements are possible for thickness, strain, composition, mosaicity, and texture.
- Throughput of XRM has improved substantially thanks to innovations in x-ray sources and optics, and rates up to 150 wafer sites/h are now possible in either XRD or XRR measurements within a typical scribe line.
- Every metrological method has a unique interaction with the surface. It should not be expected that they will all agree perfectly. However, XRM data compare well with other methods, especially for thinner films, in which the interpretation of XRM is more certain than is that of optical methods.

References

1. D. Windover, J.P. Cline, A. Henins, and M. Mendenhall, Presented at Denver X-Ray Conference, CO, 2003. Reported in NIST Office of Microelectronics Programs Report (EEEL) *NISTIR*, 7171, 2005.
2. D.K. Bowen, D. Joyce, P. Ryan, and M. Wormington, Accuracy and Repeatability of X-Ray Metrology, paper presented at ULSI 2005, NIST, Gaithersburg, MD.
3. S. Kohli, C.D. Rithner, P.K. Dorhout, A.M. Dummer, and C.S. Menoni, *Rev. Sci. Instrum.*, 76 (2005) 023906.

Index

D

E

F

G

H

L

M

N

O

P

Q

R

S

T

U

V

W

X

Y

Z